Linear Algebra

First Edition

Patrick Zulkowski

Berkeley City College

cognella ACADEMIC PUBLISHING

Bassim Hamadeh, CEO and Publisher
Kristina Stolte, Acquisitions Editor
Michelle Piehl, Project Editor
Berenice Quirino, Associate Production Editor
Miguel Macias, Senior Graphic Designer
Natalie Piccotti, Senior Marketing Manager
Kassie Graves, Director of Acquisitions and Sales
Jamie Giganti, Senior Managing Editor

Printed in the United States of America

ISBN: 978-1-5165-0255-4 (pbk) / 978-1-5165-0256-1 (br)

cognella® ACADEMIC PUBLISHING

This book is dedicated to Alleen.

Without her patience and support this work would not have been possible.

Contents

Chapter 1

Linear Systems of Equations

1.1 Why Linear Algebra?

Learning a new math subject like linear algebra can feel like wading into a relentless stream of theorems, problems, and more problems. It is very easy to get lost in the details and not see the big picture. Understanding why you are studying linear algebra-what it is used for and where it all leads-can be very helpful. So I will attempt to answer the question "Why Linear Algebra?"

For the time being, I will define linear algebra as the branch of mathematics concerned with **linear systems of equations** and their solutions. What does that mean? Let's look at an example together.

Application: Planning a Pizza Party

Sally just won a \$65 catered lunch in a contest at Pat's Pizzeria and wants to treat her friends. She has the choice of plain cheese pizza (\$10.00 per pie), vegetable pizza (\$11.00 per pie), and pepperoni pizza (\$12.00 per pie), each coming in eight slices. Sally has sixteen friends, and they'll each eat only three slices. Many of her friends prefer vegetable pizza, so she decides to order as many vegetable pizzas as pepperoni and plain cheese combined. How many pies of each type should Sally order?

Let's deconstruct the problem and approach it mathematically. First, we should identify what is **unknown**; in other words, the quantity or quantities we need to find based on the information given. In this problem, the unknowns are the number of plain cheese, vegetable, and pepperoni pies Sally should order for her guests. Writing "number of plain cheese pies" and so on over and over again is tedious and inefficient. So I'll call these unknown numbers C (for plain cheese), V (for vegetable), and P (for pepperoni). Right now, there are no restrictions on these numbers. (You could argue that we can't order fractions of a pizza or negative pizzas, but let's not split hairs (or pizzas!) just yet.) We are free to vary C, V, and P as we please, so they are examples of **variables**.

The first step has been accomplished: we've identified our unknown variables (C, P, and V). Now let's recall the given information. We know that for each plain cheese pizza Sally must spend \$10.00, for each vegetable pizza Sally must spend \$11.00, and for each pepperoni pizza Sally must spend \$12.00. The total bill has to come out to be \$65.00, so we have an **equation** that restricts the possible values for C, V, and P:

$$(\$10.00)\,C + (\$11.00)\,V + (\$12.00)\,P = \$65.00. \tag{1.1.1}$$

This isn't the only equation the word problem involves: remember that Sally's expecting sixteen guests who'll eat three slices each. This means that the total number of slices has to be exactly 3 slices per guest $\times$ 16 guests $=$ 48 slices. Each pizza has eight slices so we have yet another equation:

$$8C + 8V + 8P = 48. \tag{1.1.2}$$

We're not done finding equations yet! Since Sally's guests prefer vegetable pizza, she decided to order as many vegetable pizzas as plain cheese and pepperoni pies combined. Therefore, we must have:

$$V = C + P. \tag{1.1.3}$$

All together, our equations are:

$$\begin{aligned} (\$10.00)\,C + (\$11.00)\,V + (\$12.00)\,P &= \$65.00 \\ 8C + 8V + 8P &= 48 \\ V &= C + P. \end{aligned} \tag{1.1.4}$$

In order to plan Sally's pizza party correctly, *all* equations in Eq. (1.1.4) have to be true at the same time. This means that the number of pizzas ordered (C, V and P) must **satisfy** each equation when the actual values of C, P, and V take the place of variables: just satisfying one or two won't do! In this sense, we have a **system of equations**; i.e. we have a set of one or more equations involving the unknown variables. These equations are **linear**, which is a term I'll define in a more precise manner later. For now, just notice that our equations don't involving anything like V^2 or e^P or $\ln C$; only constants and first powers of the variables are present in our equations.

Formulating the word problem as a mathematics problem is only the first step (though an incredibly important one). It is now our job to find **solutions** to the equations. Solutions are simply values of the variables C, V, and P that make all equations true when the values replace the corresponding variables. I'll show that $C = 2, V = 3$, and $P = 1$ together give a solution to our problem. First,

$$\begin{aligned}(\$10.00)\,(C=2) + (\$11.00)\,(V=3) + (\$12.00)\,(P=1)\\ = \$20.00 + \$33.00 + \$12.00 = \$65.00\end{aligned} \tag{1.1.5}$$

so that equation is satisfied. Also,

$$8(C=2) + 8(V=3) + 8(P=1) = 16 + 24 + 8 = 48 \tag{1.1.6}$$

and

$$V = 3\ ,\ C + P = 2 + 1 = 3 \tag{1.1.7}$$

which means that $V = C + P$.

You may wonder how I managed to find these numbers. In fact, you may wonder if this is the only possible way to order the pizzas. Would something like $V = 6$ and $C = P = 3$ work? Well, one equation would be satisfied, but not all of them together. Maybe you think this is just a bad example and there is another answer out there somewhere. I promise you there isn't. If you play around with the possible values of C, V, and P you may feel that we were lucky in finding a solution to the system at all. I'll show you solutions are not always guaranteed to exist for general systems of equations.

Linear algebra is concerned with answering questions like "Does a system of linear equations have a solution?" and "If so, how many are there, and how do I find them?". In the first question, we are considering the problem of **consistency**. A system of linear equations is called consistent if there is at least one solution. Since I am able to write down a solution for our pizza party example, it involves a consistent system of linear equations. The second question deals with the issue of **uniqueness**: if *one* solution exists, could there potentially be others? It turns out that $C = 2, V = 3$, and $P = 1$ are the only possibilities for the pizza party problem, so there is a **unique solution**.

The pizza party example is a nice, down-to-earth way to start thinking about linear algebra and linear systems of equations. Solving mundane problems like ordering food for guests is just the tip of the iceberg. If the subject just dealt with finding definite solutions of linear systems of equations, it could be mastered in a few days. We'll see that exploring linear systems of equations puts us on a path to discovering many new and beautiful mathematical ideas. These ideas are applied to many real world phenomena ranging from financial equilibria to chemical reactions to the world of the subatomic. I'll show you a few more examples to convince you that linear algebra is a subject worth mastering.

Linear Algebra and Cryptography

Alice and Charlie like to pass notes to each other in class. However, Bob sits between them and often reads the messages they pass back and forth. Alice becomes tired of Bob's meddling and decides to use what she knows about math to **encrypt** her messages to Charlie. During recess, she explains to Charlie that she can assign a number to every letter of the alphabet starting at "0" and going all the way up to "25":

A	B	C	D	E	F	G	H	I	J	K	L	M	N	O	P	Q	R	S	T	U	V	W	X	Y	Z
0	1	2	3	4	5	6	7	8	9	10	11	12	13	14	15	16	17	18	19	20	21	22	23	24	25

"It's easy to turn numbers into other numbers," she explains to Charlie. "So let's agree to send messages in two-letter units, like NO and HI, which are $\begin{pmatrix} 13 \\ 14 \end{pmatrix}$ and $\begin{pmatrix} 7 \\ 8 \end{pmatrix}$ in terms of pairs of numbers. Then we can scramble the numbers (and hence the letters) when we send our messages and unscramble the numbers (and hence the letters) when we receive our messages."

"Okay," Charlie says, "but how do we do that?"

"Let's say the two-letter unit has the number representation $\begin{pmatrix} x_1 \\ x_2 \end{pmatrix}$. We can scramble it up by changing the first number to $y_1 = 2x_1 + 3x_2$ and the second number to $y_2 = 7x_1 + 8x_2$," suggests Alice.

Charlie is no slouch and he notices a problem. "If the numbers x_1 and x_2 are big, won't y_1 and y_2 be bigger than 25? What letters do they correspond to?"

Alice thinks a moment. "I have it! We can just take y_1 and y_2 **modulo** 26! In other words, we just add or subtract multiples of 26 to or from y_1 and y_2 until we reach a new pair of integers between 0 and 25."

"So NO, which is $\begin{pmatrix} 13 \\ 14 \end{pmatrix}$ becomes $\begin{pmatrix} 16 \\ 21 \end{pmatrix}$, which is QV. And HI, which is $\begin{pmatrix} 7 \\ 8 \end{pmatrix}$ becomes $\begin{pmatrix} 12 \\ 9 \end{pmatrix}$, which is MJ."

"Exactly!" says Alice. "Now we just need a way of **decrypting** our messages..."

Throughout history, people like Alice and Charlie have used various methods for sending information to each other in disguised form to fool unintended recipients like nosey Bob. **Cryptography** is the study of such methods.

Linear algebra is used in cryptography in **enciphering methods**. When Alice and Charlie defined $y_1 = 2x_1 + 3x_2$ and $y_2 = 7x_1 + 8x_2$, they were, in fact, performing a **linear transformation** (also known in cryptography as an **enciphering transformation**) on the **plain text**. The plain text is the readable message written as $\begin{pmatrix} x_1 \\ x_2 \end{pmatrix}$. The purpose of enciphering is to disguise the plain text in **cipher text** (written as $\begin{pmatrix} y_1 \\ y_2 \end{pmatrix}$). Changing a plain text message to a cipher text message is called **enciphering** or **encryption**, and the reverse process is called **deciphering** or **decryption**.

As Alice realized, the hard part is trying to determine a method of unscrambling, or deciphering, the sent messages. Alice and Charlie need some way of undoing the encryption through a **decryption transformation**. If you think of encrypting as a function of pairs of integers to pairs of integers (i.e. a rule for assigning pairs of integers so that one input pair is assigned one output pair), then decrypting is the **inverse** function.

Later in this book, we will see how linear algebra methods can be used to help solve problems similar to Alice and Charlie's decryption problem. A table of numbers of the form $\begin{pmatrix} a \\ b \end{pmatrix}$ is referred to as a **vector**. Vectors appeared in this section to represent plain text and cipher text. Vectors have a single column of numbers all contained in a set of parentheses " ( )."[1] In this example, all vectors have exactly two entries or **components**. We will encounter tables of numbers with an arbitrary number of rows and columns. These are collectively referred to as **matrices** (plural of **matrix**) of which vectors are a specific example. Working with matrices and vectors is a basic skill learned when mastering linear algebra. The next section details an example where matrices are useful in a different way.

Linear Algebra and Graph Theory

Richard is the operations manager of a package delivery company. The company uses Mercury Airways, which flies to 70 cities around the world. Richard needs to send a package from Boston to Berlin with the fewest number of connecting flights to save on money. How can he plan the itinerary of the package?

This is precisely the sort of problem that **graph theory** is well-suited to answer. You may be familiar with the graph of a function, which is a useful visualization tool. The "graph" in graph theory is a different mathematical model but is still useful for visualizing mathematics problems like Richard's delivery dilemma. A **graph** is a collection of objects represented by points called **vertices** connected by lines called **edges**. Not every pair of vertices will be connected by an edg,e typically, so a **relation** between the vertices is used to determine which vertices are connected by an edge and which are not.

[1]Sometimes square brackets " [ ] " are used instead.

Here's a visual example to help with the definition. Let's label the vertices 1 through 7. In Figure 1.1, vertices 1 and 2 are connected with an edge, but vertices 1 and 3 are not. So vertices 1 and 2 are related, but vertices 1 and 3 are not.

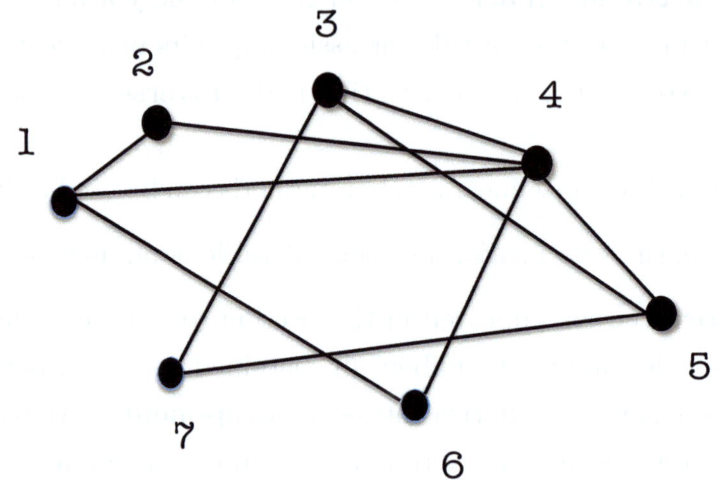

Figure 1.1: Visualization of a basic graph. Here, we have seven vertices. Some pairs are connected with an edge, other pairs are not. Those connected with an edge are **related**.

If Richard represents each of the cities serviced by Mercury Airways as dots or vertices and connects two vertices with an edge if and only if Mercury has a direct flight between them, then Richard would obtain a graph similar to the one in Figure 1.1, just with more vertices and edges. For simplicity, though, let's study the graph in the figure, since it has seven vertices whereas Richard's would have 70.

In analogy to Richard's dilemma, let's say we want to walk from vertex 1 to vertex 5 along the fewest possible edges. (This is like Richard's need to find a sequence of direct flights from Mercury-serviced cities that connects Boston to Berlin with the fewest possible connecting flights.) Notice that vertices 1 and 5 are not connected with a single edge in Figure 1.1. However, we could walk in the following sequence to get from 1 to 5: $1 - 2 - 4 - 5$. This would be an example of a 3-walk since we must cross 3 edges to complete the journey. A shorter walk would be $1 - 4 - 5$, which is a 2-walk. A walk made up of n edges is, in general, called an n-walk.

If you think about Richard's graph with its 70 vertices and a multitude of edges, you get a sense of how difficult Richard's problem can be without the aid of linear algebra, which can make the problem more manageable. In the previous example with Alice and Charlie, I introduced the idea of a **matrix**. A matrix is simply a table or array of numbers arranged in rows and columns. For a general graph, the **adjacency matrix** is a table of 1's and 0's. In row i and column j $(i \neq j)$, a 1 is written if the vertex i is connected to vertex j with an edge; a 0 is written if the vertices are not connected with an edge. We also write a 0 in the entry of the matrix corresponding to row i and column i; in other words, conventionally, a vertex is not connected to itself by an edge.

The adjacency matrix for the graph in the figure is given by

$$A = \begin{pmatrix} 0 & 1 & 0 & 1 & 0 & 1 & 0 \\ 1 & 0 & 0 & 1 & 0 & 0 & 0 \\ 0 & 0 & 0 & 1 & 1 & 0 & 1 \\ 1 & 1 & 1 & 0 & 1 & 1 & 0 \\ 0 & 0 & 1 & 1 & 0 & 0 & 1 \\ 1 & 0 & 0 & 1 & 0 & 0 & 0 \\ 0 & 0 & 1 & 0 & 1 & 0 & 0 \end{pmatrix}. \tag{1.1.8}$$

Here is an interesting theorem that I'll prove for you later in the book: the (i, j) entry (the number in row i and column j) of A^n is precisely the number of n-walks connecting vertices i and j. Of course, we don't yet know how to compute A^n for positive integer n just yet or even what it means, so I'll write down A^2 and A^3 for you for the graph provided in Figure 1.1:

$$A^2 = \begin{pmatrix} 3 & 1 & 1 & 2 & 1 & 1 & 0 \\ 1 & 2 & 1 & 1 & 1 & 2 & 0 \\ 1 & 1 & 3 & 1 & 2 & 1 & 1 \\ 2 & 1 & 1 & 5 & 1 & 1 & 2 \\ 1 & 1 & 2 & 1 & 3 & 1 & 1 \\ 1 & 2 & 1 & 1 & 1 & 2 & 0 \\ 0 & 0 & 1 & 2 & 1 & 0 & 2 \end{pmatrix}, \quad A^3 = \begin{pmatrix} 4 & 5 & 3 & 7 & 3 & 5 & 2 \\ 5 & 2 & 2 & 7 & 2 & 2 & 2 \\ 3 & 2 & 4 & 8 & 5 & 2 & 5 \\ 7 & 7 & 8 & 6 & 8 & 7 & 2 \\ 3 & 2 & 5 & 8 & 4 & 2 & 5 \\ 5 & 2 & 2 & 7 & 2 & 2 & 2 \\ 2 & 2 & 5 & 2 & 5 & 2 & 2 \end{pmatrix}. \tag{1.1.9}$$

The $(1,5)$ and $(5,1)$ entries are 0, which means that there is no edge connecting the two vertices and, hence, no 1-walks. If you believe the theorem I just stated, then the $(1,5)$ and $(5,1)$ entries of A^2 (which have to be the same) should tell us the number of 2-walks between the vertices. That entry of A^2 is 1, meaning there is exactly one 2-walk between the vertices, namely $1-4-5$.

This simple example should give you an idea of the power of linear algebra. If Richard builds the adjacency matrix of the graph representing the flight paths of Mercury Airlines and can calculate **matrix powers**, he will be able to find the length of the shortest route between Boston and Berlin by computing $A, A^2, A^3, \ \ldots$ until he reaches a matrix power A^k where the Boston-Berlin entry is nonzero.

Graph theory is a beautiful subject that has found applications in other areas, such as sociology, physics, and social media. I will return to this application of linear algebra later in my coverage of matrix powers, **symmetric matrices**, and **matrix diagonalization**.

Linear Algebra and Medical Imaging

Wilhelm Röntgen kicked off a medical revolution in 1895 when he discovered X-rays. X-rays are light waves that may be used to study the inside of the human body without surgery. Medical technology has evolved since Röntgen's time. Computer-assisted tomography scans are commonly used in many medical practices, transmission tomography being one example.

In transmission tomography, radiation (normally in the form of X-rays) passes through an object (a person, for example) along straight line paths. The intensity of the emitted radiation is known and the intensity of the radiation after it passes through the object is measured and recorded. The distribution of mass throughout the object alters the intensity of the radiation coming out of the object: some of it gets absorbed on its journey from emitter to detector.

To get a sense of what is happening, think of a marathon with hundreds of participants. Eventually all of the runners will make it past the finish line assuming the street stays clear. But suppose instead that the fans and spectators spill out into the street: some runners may deftly maneuver to the finish line, but most will get held up for autographs or pats on the back.

The term "tomography" arises from the Greek word for "slice" or "part." The scan essentially focuses on one slice of the object at a time by sending radiation beams through the slice and measuring the output. The resulting image is made up of a large number of little squares called **pixels**. When the radiation beams are measured, a computer is used to compute the pixel densities based on these measurements. The pixel densities correspond to mass densities in the scanned object; a lighter pixel corresponds to a spot with less mass, a darker pixel corresponds to a spot with more mass. If we label the pixel densities $x_1, x_2, \ldots$ and so on, the computer has to solve a linear system of equations for these unknowns in order to display an image of the object's interior.

In fact, for every radiation beam used, there will be one linear equation, so if M beams are used to scan the object, M linear equations need to be solved simultaneously. Typically, M will be larger than N, the number of pixel densities, which is also the number of unknowns. When a linear system has more equations than unknowns, the system is **overdetermined**. In such situations, it is possible that the system is inconsistent. Remember, inconsistency means no solution. If that happens with the linear system of equations coming from a CAT scan, how can linear algebra be used to find the pixel densities and image the interior?

There won't be values that satisfy all equations, but there may be a set of values that comes as close as possible. That is, it may be possible to find values $x_1, x_2, \ldots$ that, in some sense, *approximately* solves the system in the best possible way. I'll formalize this idea when we study the **least squares method** later in the book. The least squares method is based on the theory of **inner product spaces**, which also has applications in the theory of differential equations and quantum physics.

1.2 Matrices, Vectors, and All That.

In this portion of the book, I introduce the terminology needed to start thinking about linear systems and the methods for finding their solution sets. Recall Sally's pizza party problem. I introduced special symbols C, V, and P for the number of different pizza types Sally could order for her friends. They represent **variables**.

Definition: A **variable** is a symbol representing a quantity without a fixed value.

You are probably accustomed to using x and y to represent variables. Choosing variable names from the Greek and Latin alphabets is a long tradition in mathematics. However, there is an immediate problem with choosing variable names from a fixed alphabet. Think about the Latin alphabet with letters A, B , ..., Z. If we decide to name our variables with Latin letters, we are limited to just 26 possibilities. Also, if we do have need for a large number of variables, it can be a bit confusing using so many different letters to represent various quantities. You could double the number of possible variable names by including lower case letters, but that just adds to the confusion, and then we are limited to just 52 possible names.

You may wonder if mathematicians ever realistically need to consider a large number of variables. Sally's pizza party problem only involved three variables after all. But think back to the medical imaging application. A computer reconstructs the cross section of a patient using X-ray radiation by solving a linear system of equations (at least approximately). The interior of the patient can be pictured with N pixels arrayed in a grid, much the same way an image is made on a computer or television screen. The shade or density of each pixel is an unknown in a large system of equations. To get a reasonably sharp resolution, let's pick an image size of 256×256 square pixels. In other words, our picture will be made up of a grid of pixels of length 256 on a side. Then, there are $256^2 = 65,536$ total pixels and, hence, $65,536$ total unknowns!

Clearly, this example illustrates how it is inefficient and/or impossible to label variables using letters from a single alphabet. We obviously need some alternative notation. Since many linear algebra problems are done on digital computers where it is natural to use lists and arrays of integers, it makes a lot of sense to represent the variables of our problem, no matter how many, as

$$x_1, x_2, \ldots, x_n \tag{1.2.1}$$

where the subscript labels the particular quantity of interest. There is no end to the positive integers, so we can represent any set of variables in a linear algebra problem this way. This convention is very convenient to use when linear algebra problems are solved on a computer.

In Sally's pizza party problem, we could make the following relabeling: $C = x_1, V = x_2$, and $P = x_3$. That way, the equations of the problem can be written as

$$\begin{aligned} 10x_1 + 11x_2 + 12x_3 &= 65 \\ 8x_1 + 8x_2 + 8x_3 &= 48 \\ x_2 &= x_1 + x_3. \end{aligned} \tag{1.2.2}$$

This is an example of a **linear system of equations**.

Definition: A **linear system of equations** is a collection of one or more linear equations in the same set of variables $x_1, \ldots, x_n$. A **linear equation** in variables $x_1, \ldots, x_n$ is an equation of the form

$$a_1x_1 + a_2x_2 + \ldots a_nx_n = b \tag{1.2.3}$$

where $a_1, \ldots, a_n$ are fixed constants called **coefficients**.

I will always assume that $a_1, \ldots, a_n$ and the fixed constant b are real or complex numbers. I'll mostly focus on the case where they are strictly real numbers, but I will have occasion to consider the complex case later in the book, especially when discussing linear systems of differential equations and inner product spaces.

Notation: In this book, I'll use $\mathbb{R}$ to indicate the set of all real numbers and $\mathbb{C}$ to indicate the set of all complex numbers. I also define $i = \sqrt{-1}$.

The equation written in (1.2.3) is the most general *single* linear equation in n variables. The subscript on a coefficient tells us which x the coefficient is paired with. For instance, a_2 must be the coefficient of x_2. To write down the most general *system* of linear equations, another number needs to be included in the subscript of each of the coefficients:

$$\begin{aligned} a_{11}x_1 + \cdots + a_{1n}x_n &= b_1 \\ &\vdots \\ a_{m1}x_1 + \cdots + a_{mn}x_n &= b_m \end{aligned} \tag{1.2.4}$$

Here, the first number in a_{ij} (in blue), which is pronounced " a sub i j, " tells us which equation coefficient a_{ij} belongs to; the second number in a_{ij} (in red) tells us which variable we should pair the coefficient with. You can think of the subscript pair (i, j) in a_{ij} as representing the "address" of where a_{ij} lives in the general linear system.

The focus of the first part of our course is finding all possible **solutions** to (1.2.4), if they exist.

Definition: An ordered list of n numbers $(s_1, \ldots, s_n)$ (called an n-tuple) is a **solution** to the general system of linear equations (1.2.4) if substituting $x_1 = s_1, \ldots, x_n = s_n$ into *every* equation makes *every* equation true simultaneously.

Example: Direct substitution shows that $x_1 = 2, x_2 = 3, x_3 = 1$ solves (1.2.2).

Sometimes a linear system of equations will have more than one solution. In fact, it necessarily has an infinite number of solutions if this is the case. Set notation provides a way to describe **solution sets** of linear systems of equations.

For our purposes, any set S can be written as

$$S = \{\text{element type} \mid \text{element conditions}\}. \tag{1.2.5}$$

Here, the pair of braces $\{\}$ indicates that we are dealing with a set and can be read "the set of ... ". The vertical line $|$ separates the description of the type of our set elements from the conditions on these elements and can be read as "such that."

Example: I'll write $\mathbb{Q}$, the set of rational numbers, explicitly using set notation. One way to do so would be

$$\mathbb{Q} = \left\{ r \in \mathbb{R} \middle| r = \frac{m}{n} \ , \ m, n \in \mathbb{Z} \ , \ n \neq 0 \right\}. \tag{1.2.6}$$

Here, the collection of symbols $r \in \mathbb{R}$ before $|$ indicates that the elements are real numbers: the symbol $\in$ means "element of" or "belongs to." After the $|$, we find our conditions: r has to be the ratio of m and n, m and n are integers $(m, n \in \mathbb{Z})$, and finally, $n \neq 0$, since we can never divide by 0. All together, we would read this set as "the set of all real numbers r such that r is the ratio of m and n, where m and n are integers and n is not zero."

Using this notation, we see that a general solution set S of real solutions can be written as

$$S = \{(s_1, \ldots, s_n) \in \mathbb{R}^n | a_{11}s_1 + \cdots + a_{1n}s_n = b_1, \ldots\}$$

where I have colored the s values to emphasize that the corresponding variables are being evaluated with these values.

Definition: The set of all n-tuples [2] of real numbers is called the **Euclidean space** of dimension n or n-dimensional Euclidean space. This set is denoted $\mathbb{R}^n$.

Note: I defined S to be the set of real solutions to our linear system of equations. When we want to allow for complex solutions, we simply change $\mathbb{R}^n$ in the above definition to $\mathbb{C}^n$, the set of all n-tuples of *complex* numbers.

You probably have seen $\mathbb{R}^2$ before, though you may have called it the **Cartesian plane** in other classes. In algebra, the Cartesian plane is represented visually by drawing two real number lines perpendicular to each other. These lines are called axes. Here, we are thinking of the plane as made up of points with coordinates (x_1, x_2); hence the Cartesian plane is really $\mathbb{R}^2$, the set of all ordered pairs (2-tuples) of real numbers.

As I mentioned in the introduction, it is possible that a linear system of equations could have no solution (i.e. be inconsistent). The CAT scan example is a real world application where this occurs. If there are no solutions to the given system of linear equations, we'll write $S = \emptyset$ to indicate that the set is **empty**. Sometimes this is also written as $\{\}$ to indicate a set with nothing inside of it.

Our near-term goal is to systematically determine the solution set of a given system of linear equations. Broadly speaking, I'll show you how to take our original system and use mathematically valid operations on the equations to transform it into a new system of equations that has precisely the same solution set. The new system will be in a form where the

[2] Remember that an n-tuple is an ordered list of n elements.

solution set is easier to determine.

Definition: Two linear systems of equations are called **equivalent** if they share precisely the same solution set. For instance, it would be nice to cancel off the common factor of 8 on both sides of the second equation in (1.2.2). The resulting system of equations

$$\begin{aligned} 10x_1 + 11x_2 + 12x_3 &= 65 \\ x_1 + x_2 + x_3 &= 6 \\ x_2 &= x_1 + x_3. \end{aligned} \tag{1.2.7}$$

is equivalent to the original pizza party system, since canceling off the common factor of 8 cannot possibly change the solution set. You should verify on your own that the two systems do indeed share the solution $x_1 = 2, x_2 = 3, x_3 = 1$.

Before I go on to describe the algorithm allowing us to find the solution set S for any given linear system of equations, allow me to clean up our notation a little.

If you think about it, a general linear system is completely determined by the coefficients a_{ij} and the b_i values. Given these quantities, we can completely reconstruct the original linear system much the same way that box and line scores allow a baseball fan to reconstruct the events of an individual game. So allow me to organize the a_{ij}'s and the b_j's into rectangular tables or arrays called **matrices**.

Definition: The array

$$\begin{pmatrix} a_{11} & \dots & a_{1n} \\ \vdots & \vdots & \vdots \\ a_{m1} & \dots & a_{mn} \end{pmatrix} \tag{1.2.8}$$

is called the **coefficient matrix** of the general linear system (1.2.4). The dots represent coefficients not explicitly written but understood to be present in the matrix. Any matrix has a specific size which is determined by the number of rows and the number of columns.

Definition: The **size** or **dimension** of a matrix with m rows and n columns is $m \times n$ (read "m by n").

A special type of matrix has only a single column; i.e. the size is $m \times 1$ for some m. At least for now, this is how I'll define a **vector**.

Definition: A **vector** or **column vector** is an $m \times 1$ matrix.

We can arrange the b_i values of our linear system into a column vector called $\vec{b}$:

$$\vec{b} = \begin{pmatrix} b_1 \\ \vdots \\ b_m \end{pmatrix}. \tag{1.2.9}$$

The overhead arrow notation for a vector $\vec{v} = \begin{pmatrix} v_1 \\ \vdots \\ v_m \end{pmatrix}$ is supposed to emphasize its geometric interpretation as an arrow from the origin of $\mathbb{R}^n$ to a point with coordinates given by the n-tuple $(v_1, \ldots, v_n)$. Don't worry about this now; I'll have much more to say about this later.

The general linear system Eq. (1.2.4) can be completely specified by the **augmented matrix**

$$\left(\begin{array}{ccc|c} a_{11} & \ldots & a_{1n} & b_1 \\ \vdots & \vdots & \vdots & \vdots \\ a_{m1} & \ldots & a_{mn} & b_m \end{array}\right), \tag{1.2.10}$$

which I will represent as $(A|\vec{b})$ to emphasize that it is made up of the coefficient matrix A augmented by the vector $\vec{b}$ of b_j values. I will use the vertical line to **partition** augmented matrices throughout the book to distinguish matrices representing linear systems from plain old matrices not necessarily affiliated with a linear system. [3]

[3]Sometimes, as in the case of an augmented matrix, we need to distinguish separate pieces of matrix. These pieces are usually smaller matrices, which we call **submatrices**. **Partitions** are imaginary lines superimposed on a matrix to indicate to the reader a subdivision of a matrix into smaller submatrices. For the augmented matrix considered above, one line separates the last column from the others: the submatrix to the left of the partition is the coefficient matrix A, and the submatrix to the right is the vector $\vec{b}$.

1.3 A Few Simple Linear Systems

I am now ready to describe a method known as **Gauss-Jordan elimination** that allows us to find solutions to linear systems of equations (if they exist) using operations on the corresponding augmented matrix.

First, I'll describe the three basic operations that we may use on a linear system. These operations leave the solution set intact: given a system of linear equations, these operations produce **equivalent systems**. Hopefully, it's clear that the order of equations in a linear system doesn't really matter when it comes to the solution set: if we switch any pair of equations in the list of linear equations, this can't affect the answer.

Example: The systems

$$\begin{aligned} 3x_1 + 6x_2 &= -3 \\ 5x_1 + 7x_2 &= 10 \end{aligned} \quad \text{and} \quad \begin{aligned} 5x_1 + 7x_2 &= 10 \\ 3x_1 + 6x_2 &= -3 \end{aligned} \tag{1.3.1}$$

have precisely the same solutions; i.e., they are equivalent. I will call this **operation** on equations the **interchange operation**.

Recall from basic algebra that if $c \neq 0$, then $ax + b = y$ has the same math content (same solutions) as $c\,(ax + b) = cy$. In other words, we may multiply any single equation we wish by a nonzero number on both sides. The number has to be nonzero, otherwise we have $0 = 0$, which tells us nothing, and we irrevocably change the system by losing an equation.

Example: The systems

$$\begin{aligned} 3x_1 + 6x_2 &= -3 \\ 5x_1 + 7x_2 &= 10 \end{aligned} \quad \text{and} \quad \begin{aligned} \tfrac{1}{3}(3x_1 + 6x_2) &= \tfrac{1}{3}(-3) \\ \tfrac{1}{5}(5x_1 + 7x_2) &= \tfrac{1}{5}(10) \end{aligned} \quad \text{or} \quad \begin{aligned} x_1 + 2x_2 &= -1 \\ x_1 + \tfrac{7}{5}x_2 &= 2 \end{aligned} \tag{1.3.2}$$

are equivalent. I will call this **operation** on equations the **scaling operation**.

The third operation requires some explanation. In short, we are allowed to add to any equation the multiple of another equation without altering the solution set. Let's look at an example to see how this works. Let me start with the system

$$\begin{aligned} 3x_1 + 6x_2 &= -3 \\ 5x_1 + 7x_2 &= 10, \end{aligned} \tag{1.3.3}$$

and let me assume the pair (s_1, s_2) is a solution to this system. (I will leave the case of inconsistent systems for the reader to ponder.) Here's one way of thinking about (s_1, s_2): s_1 and s_2 allow me to write -3 as $3s_1 + 6s_2$. This follows from the first equation in the linear system. Additionally, s_1 and s_2 allow me to write 10 as $5s_1 + 7s_2$.

So, if I add any multiple of -3 to 10, we have

$$5s_1 + 7s_2 + c\,(-3) = 10 + c\,(-3). \tag{1.3.4}$$

But, we know that $-3 = 3s_1 + 6s_2$, so I can rewrite the equation above as

$$5s_1 + 7s_2 + c\,(3s_1 + 6s_2) = 10 + c\,(-3) \tag{1.3.5}$$

or

$$(5 + 3c)\,s_1 + (7 + 6c)\,s_2 = 10 - 3c \tag{1.3.6}$$

So you see that s_1 and s_2 are also solutions to the new linear system

$$\begin{aligned} 3x_1 + 6x_2 &= -3 \\ (5 + 3c)\,x_1 + (7 + 6c)\,x_2 &= 10 - 3c \end{aligned} \tag{1.3.7}$$

where c is an arbitrary real number.

With this example, I hope I have convinced you that adding a constant multiple of one equation to another does not alter the solution set. One example does not constitute a proof, of course.

This operation is referred to as the **replacement operation**. The three equation operations (**replacement**, **scaling** and **interchange**) carry over directly to operations on the augmented matrix. I list them here for clarity:

- (Interchange) Interchanging two rows of an augmented matrix leads to an equivalent augmented matrix.
- (Scaling) Multiplying all entries in a row of an augmented matrix by a nonzero constant leads to an equivalent augmented matrix.
- (Replacement) Replacement of one row of an augmented matrix by the sum of itself and a multiple of another row leads to an equivalent augmented matrix.

To be clear, two augmented matrices are **equivalent** if the corresponding linear systems are equivalent. Just to drive home the point, only these operations can be performed on an augmented matrix without altering the solution set of the original system. In other words, the **elementary row operations** listed above generate equivalent augmented matrices.

Let's solve the system Eq. (1.3.3) using these elementary row operations. First, we form the augmented matrix:

$$\left(\begin{array}{cc|c} 3 & 6 & -3 \\ 5 & 7 & 10 \end{array}\right). \tag{1.3.8}$$

Our goal will be to find a simpler, equivalent augmented matrix with an easy-to-determine solution set. I'll be more systematic about this shortly, but let's see how far we can get with this example just playing around with the elementary row operations.

Notice that the first row has components that share a common factor of 3. I can use a scaling operation to remove this common factor:

$$\left(\begin{array}{cc|c} 3 & 6 & -3 \\ 5 & 7 & 10 \end{array}\right) \begin{array}{c} r_1 \to \frac{1}{3}r_1 \\ r_2 \to r_2 \end{array} \left(\begin{array}{cc|c} 1 & 2 & -1 \\ 5 & 7 & 10 \end{array}\right). \tag{1.3.9}$$

The notation $r_1 \to \frac{1}{3}r_1$ means I am replacing row 1 with row 1 scaled by 1/3, and $r_2 \to r_2$ means that I am leaving row 2 alone. For at least the beginning of the book, I will be explicit about the row operations used on a matrix in this way. I would encourage you to do the same until you have enough experience with row reduction.

While this operation does simplify the matrix, it is not yet simple enough: we can't yet write down the solution set by inspecting this new (equivalent) matrix. Remember that we are free to add to any row in the matrix a multiple of another. This is the replacement operation. Notice that if I add -5 times row 1 to row 2, the 5 in row 2 will become a 0. If you think about the corresponding linear system, this means that x_1 will be eliminated[4] from the second equation, leaving only x_2. Let's do this!

$$\left(\begin{array}{cc|c} 1 & 2 & -1 \\ 5 & 7 & 10 \end{array}\right) \xrightarrow[r_2 \to r_2 + (-5)r_1]{r_1 \to r_1} \left(\begin{array}{cc|c} 1 & 2 & -1 \\ 0 & -3 & 15 \end{array}\right). \tag{1.3.10}$$

This is certainly progress, but I can do better: let me scale the second row by $-1/3$ to get rid of the pesky common factor of -3:

$$\left(\begin{array}{cc|c} 1 & 2 & -1 \\ 0 & -3 & 15 \end{array}\right) \xrightarrow[r_2 \to \left(-\frac{1}{3}\right) r_2]{r_1 \to r_1} \left(\begin{array}{cc|c} 1 & 2 & -1 \\ 0 & 1 & -5 \end{array}\right). \tag{1.3.11}$$

I managed to use the replacement operation to remove the x_1 variable from the second equation. I accomplished this by using the 1 (the coefficient of x_1 in equation 1) to eliminate the 5 (the coefficient of x_1 in equation 2).

It is tempting to repeat this process by using the 1 in the second row to eliminate the 2 above it. In terms of equations, this means that I am eliminating x_2 from equation 1. The replacement operation allows me to do this, so let me write:

$$\left(\begin{array}{cc|c} 1 & 2 & -1 \\ 0 & 1 & -5 \end{array}\right) \xrightarrow[r_2 \to r_2]{r_1 \to r_1 + (-2)r_2} \left(\begin{array}{cc|c} 1 & 0 & 9 \\ 0 & 1 & -5 \end{array}\right). \tag{1.3.12}$$

By playing this eliminate-the-variable game, we come to the equivalent augmented matrix $\left(\begin{array}{cc|c} 1 & 0 & 9 \\ 0 & 1 & -5 \end{array}\right)$ by way of elementary row operations. This augmented matrix must have solution set identical to S for our original system Eq. (1.3.3)[5]. In terms of linear equations, the new augmented matrix describes the linear system

$$x_1 = 9\ ,\ x_2 = -5. \tag{1.3.13}$$

[4]Hence the name "Gauss-Jordan elimination."

[5]Here and throughout the text, Eq. stands for equation

The solution set is then obvious! S must be $\{(9, -5)\}$ for *both* systems.

Before I move on to the algorithm of Gauss-Jordan elimination, let's look at another example. Recall the equations I derived for Sally's pizza party problem:

$$\begin{aligned} 10x_1 + 11x_2 + 12x_3 &= 65 \\ 8x_1 + 8x_2 + 8x_3 &= 48 \\ x_2 &= x_1 + x_3. \end{aligned} \tag{1.3.14}$$

I can write Eq. (1.3.14) in terms of its augmented matrix:

$$\left(\begin{array}{ccc|c} 10 & 11 & 12 & 65 \\ 8 & 8 & 8 & 48 \\ -1 & 1 & -1 & 0 \end{array}\right). \tag{1.3.15}$$

Just like our last example, let's see if we can play the elimination game to eliminate as many variables from the equations as possible. In terms of the augmented matrix, this means using elementary row operations to change entries of the coefficient matrix (the part to the left of the partition) to zeros.

Starting from the left and working our way right, we see that we can use the following options:

1. We can use the 10 in the first entry of the first row to eliminate the 8 and the -1.
2. We can use the 8 to eliminate the -1 and the 10.
3. We can use the -1 to eliminate the 8 and the 10.

All of these options are possible thanks to the elementary row operations. The easiest route would be to use the -1 in the last row to eliminate the entries above, so I'll do this:

$$\left(\begin{array}{ccc|c} 10 & 11 & 12 & 65 \\ 8 & 8 & 8 & 48 \\ -1 & 1 & -1 & 0 \end{array}\right) \begin{array}{c} r_1 \to r_1 + 10r_3 \\ r_2 \to r_2 + 8r_1 \\ r_3 \to r_3 \end{array} \left(\begin{array}{ccc|c} 0 & 21 & 2 & 65 \\ 0 & 16 & 0 & 48 \\ -1 & 1 & -1 & 0 \end{array}\right) \tag{1.3.16}$$

If I emulate what I did in the last example as closely as possible, I can use the interchange operation to move the two top rows, which now have zeros for their first entries to the bottom. I also can change the -1 into a $+1$ by scaling:

$$\left(\begin{array}{ccc|c} 0 & 21 & 2 & 65 \\ 0 & 16 & 0 & 48 \\ -1 & 1 & -1 & 0 \end{array}\right) \begin{array}{c} r_3 \to (-1)r_3 \\ r_1 \leftrightarrow r_3 \\ r_2 \leftrightarrow r_3 \end{array} \left(\begin{array}{ccc|c} 1 & -1 & 1 & 0 \\ 0 & 21 & 2 & 65 \\ 0 & 16 & 0 & 48 \end{array}\right). \tag{1.3.17}$$

It is tempting now to use the -1 in the first row to eliminate the 21 and the 16 below, but there is a problem if we do this. Even though there is nothing wrong with applying elementary row operations to achieve this, it would be counter productive. Because of the the 1 that now sits in the first entry of the first row, elementary row operations using row 1 to eliminate the 21 and the 16 in rows 2 and 3 would yield nonzero first entries for the second and third rows. In other words, we would bring the variable x_1 back into equations 2 and 3 after our hard work eliminated it in the first step!

For this reason, we should ignore the first row and just focus on eliminating x_2 or x_3 from the remaining equations. The numbers 21 and 16 are awkward to work with, but notice that the nonzero entries of the third row have a common factor of 16. I'll scale row 3 by 1/16:

$$\left(\begin{array}{ccc|c} 1 & -1 & 1 & 0 \\ 0 & 21 & 2 & 65 \\ 0 & 16 & 0 & 48 \end{array}\right) \begin{array}{c} r_1 \to r_1 \\ r_2 \to r_2 \\ r_3 \to \left(\frac{1}{16}\right) r_3 \end{array} \left(\begin{array}{ccc|c} 1 & -1 & 1 & 0 \\ 0 & 21 & 2 & 65 \\ 0 & 1 & 0 & 3 \end{array}\right). \tag{1.3.18}$$

Now I'll use the 1 in row 3 to eliminate the 21 above it. While I am at it, I'll also interchange the last two rows so that the 1 is in the second entry of the second row, just like in our last example:

$$\left(\begin{array}{ccc|c} 1 & -1 & 1 & 0 \\ 0 & 21 & 2 & 65 \\ 0 & 1 & 0 & 3 \end{array}\right) \begin{array}{c} r_1 \to r_1 \\ r_2 \to r_2 + (-21)r_3 \\ r_3 \leftrightarrow r_2 \end{array} \left(\begin{array}{ccc|c} 1 & -1 & 1 & 0 \\ 0 & 1 & 0 & 3 \\ 0 & 0 & 2 & 2 \end{array}\right). \tag{1.3.19}$$

The nonzero entries of the last row now have a common factor of 2, and it is too tempting not to use a scaling operation to cancel that common factor:

$$\left(\begin{array}{ccc|c} 1 & -1 & 1 & 0 \\ 0 & 1 & 0 & 3 \\ 0 & 0 & 2 & 2 \end{array}\right) \begin{array}{c} r_1 \to r_1 \\ r_2 \to r_2 \\ r_3 \to \frac{1}{2}r_3 \end{array} \left(\begin{array}{ccc|c} 1 & -1 & 1 & 0 \\ 0 & 1 & 0 & 3 \\ 0 & 0 & 1 & 1 \end{array}\right). \tag{1.3.20}$$

This new equivalent augmented matrix represents a linear system of equations with a solution set that is precisely the same as the solution set to the original pizza party system. The last row, once translated back into a linear equation, tells us that $x_3 = 1$; the second row tells us that $x_2 = 3$. Now, we could use this information to determine x_1 from the equation corresponding to row 1, but allow me to continue to use row operations to eliminate x_2 and x_3 from the first equation. That way, we can simply read off the solution from the last column.

Moving now from right to left, you see that subtracting row 3 from row 1 will eliminate x_3 from the first equation and adding row 2 to row 1 will eliminate x_2:

$$\left(\begin{array}{ccc|c} 1 & -1 & 1 & 0 \\ 0 & 1 & 0 & 3 \\ 0 & 0 & 1 & 1 \end{array}\right) \begin{array}{c} r_1 \to r_1 + r_2 - r_3 \\ r_2 \to r_2 \\ r_3 \to r_3 \end{array} \left(\begin{array}{ccc|c} 1 & 0 & 0 & 2 \\ 0 & 1 & 0 & 3 \\ 0 & 0 & 1 & 1 \end{array}\right). \tag{1.3.21}$$

The augmented matrix in this last step represents the linear system $x_1 = 2$, $x_2 = 3$ and $x_3 = 1$. This is how I calculated the solution to Sally's pizza party problem!

Example summary: Starting with the augmented matrix representing the original system of equations, I used elementary row operations to obtain a simple equivalent augmented matrix from which I could easily deduce the solution set.

Of course, "simple" is a vague and relative term. Before I define precisely what I mean by a simple equivalent matrix, let's look at two more examples.

Example: Find the solution set S for the system of equations described by $x_1+x_2+x_3+x_4 = 0$ and $x_1 - x_2 + x_3 - x_4 = 0$.

To start, I'll construct the augmented matrix corresponding to this system:

$$\left(\begin{array}{cccc|c} 1 & 1 & 1 & 1 & 0 \\ 1 & -1 & 1 & -1 & 0 \end{array}\right). \tag{1.3.22}$$

I'll emulate our elimination strategy from the last two examples here. Starting at the very left of the matrix, I'll use the 1 at the top of the first column to eliminate the 1 below, thereby eliminating the variable x_1 from the second equation:

$$\left(\begin{array}{cccc|c} 1 & 1 & 1 & 1 & 0 \\ 1 & -1 & 1 & -1 & 0 \end{array}\right) \begin{array}{c} r_1 \to r_1 \\ \overrightarrow{r_2 \to r_2 - r_1} \end{array} \left(\begin{array}{cccc|c} 1 & 1 & 1 & 1 & 0 \\ 0 & -2 & 0 & -2 & 0 \end{array}\right). \tag{1.3.23}$$

Canceling the common factor of -2 from the nonzero entries of row 2 and using row 2 to eliminate x_2 from equation 1, we have

$$\left(\begin{array}{cccc|c} 1 & 1 & 1 & 1 & 0 \\ 0 & -2 & 0 & -2 & 0 \end{array}\right) \begin{array}{c} r_2 \to \left(-\frac{1}{2}\right) r_2 \\ r_1 \to r_1 - r_2 \end{array} \left(\begin{array}{cccc|c} 1 & 0 & 1 & 0 & 0 \\ 0 & 1 & 0 & 1 & 0 \end{array}\right). \tag{1.3.24}$$

This is as far as we can go eliminating variables from the equations. The first row tells us that $x_1 + x_3 = 0$ or $x_1 = -x_3$ and the second row tells us that $x_2 = -x_4$.

In this example we don't have a unique answer, though our system is consistent. In fact, we have an infinite family of solutions since I could choose any value $t \in \mathbb{R}$ I wish for x_3 and any value $s \in \mathbb{R}$ I wish for x_4 and $(-t, -s, t, s)$ would be a solution. Therefore

$$S = \{(-t, -s, t, s) \in \mathbb{R}^4 | t, s \in \mathbb{R}\}. \tag{1.3.25}$$

Thus we could describe the solution set **parametrically** in this way; i.e. the solutions are expressed in terms of **parameters** that are free to take on any real value. I'll have more to say about parameters and infinite solution sets later.

Example: Find the solution set S for the system of equations described by $x_1+x_2+x_3 = 0$, $x_1 + x_2 + x_3 = 1$, and $2x_1 + 2x_2 + 2x_3 = 0$.

Starting with the augmented matrix

$$\left(\begin{array}{ccc|c} 1 & 1 & 1 & 0 \\ 1 & 1 & 1 & 1 \\ 2 & 2 & 2 & 0 \end{array}\right), \tag{1.3.26}$$

I can eliminate x_1 from the second and third equations using row 1:

$$\left(\begin{array}{ccc|c} 1 & 1 & 1 & 0 \\ 1 & 1 & 1 & 1 \\ 2 & 2 & 2 & 0 \end{array}\right) \begin{array}{c} r_1 \to r_1 \\ r_2 \to r_2 - r_1 \\ r_3 \to r_3 - 2r_1 \end{array} \left(\begin{array}{ccc|c} 1 & 1 & 1 & 0 \\ 0 & 0 & 0 & 1 \\ 0 & 0 & 0 & 0 \end{array}\right). \tag{1.3.27}$$

It's easy to see that I've eliminated all the variables I can. However, notice that this equivalent augmented matrix has a quirky feature: if we translate back to equations, the second equation tells us that $0x_1+0x_2+0x_3 = 0 = 1$. This cannot be true, so we conclude naturally that the system is inconsistent. Simply put, $S = \emptyset$.

I want to make two important observations based on our examples so far. First, I can now make precise what I mean by a "simple" equivalent matrix. Notice that all of the final matrices produced using the elementary row operations have the following features in common:

1. If the matrix possesses **zero rows** (i.e. rows with entires all 0), they appear at the bottom of the matrix below any **nonzero rows**.

2. In a nonzero row, the **leading entry** (i.e. the leftmost nonzero entry) is to the right of the leading entry of the row above it.

3. All leading entries have value 1.

4. A leading entry is the only nonzero entry in its column.

Any matrix possessing these properties (even if it does not correspond directly to a linear system of equations) is said to be in **reduced row echelon form**. "Echelon" is derived from the Old French word for the rung of a ladder. In this context, "echelon" means the matrix possesses a step-like pattern. For instance, consider the matrix produced from elementary row operations from our last example:

$$\begin{pmatrix} 1 & 1 & 1 & 0 \\ \hline 0 & 0 & 0 & 1 \\ \hline 0 & 0 & 0 & 0 \end{pmatrix}. \tag{1.3.28}$$

Here, I have removed the partition in the augmented matrix for clarity.

When I eliminated variables from left to right in our examples, I was really using elementary row operations to obtain equivalent augmented matrices in **row echelon form**. A matrix in row echelon form only satisfies the first two properties in the list above. You should go back through the examples and verify this for yourself.

Earlier when I wrote that our goal was to obtain a simple equivalent augmented matrix from which the solution set S could be easily decided, I really meant that our goal was to find an augmented matrix in reduced row echelon form. While elementary row operations can put any given nonzero matrix into an echelon form, such an echelon form is not unique. You can always scale some nonzero row to obtain an equivalent but different matrix in echelon form.

The advantage of defining the **reduced row echelon form** of a matrix is seen in the following theorem:

Theorem: Any nonzero matrix is row equivalent to a *unique* matrix of equal size in reduced row echelon form.

The existence part of the proof of this theorem (i.e. showing that such a matrix exists) comes from the Gauss-Jordan algorithm that I will describe in just a moment. The uniqueness part is left to the reader as an exercise.

The second observation I wish to make is the following: in all of the examples considered above, we only had three possibilities for our solution set S. I gave you an example of an inconsistent system in which $S = \emptyset$. For the consistent system examples, there was either a unique solution, or possesses an infinite family of solutions. Once you understand matrices and matrix operations, I'll be able to prove that these are the only possibilities!

Theorem: Given a linear system of equations, the solution set S is either empty ($S = \emptyset$), possesses a single element or an infinite number of elements.

While I cannot easily prove this theorem yet, it is not hard to visualize why this must be so. Consider again the linear system

$$\begin{aligned} 3x_1 + 6x_2 &= -3 \\ 5x_1 + 7x_2 &= 10. \end{aligned} \tag{1.3.29}$$

Both equations can be graphed separately on the Cartesian plane: by plotting all points (x_1, x_2) that satisfy $3x_1 + 6x_2 = -3$, we obtain the graph of the equation. This graph turns out to be a line. Similarly, the graph of $5x_1 + 7x_2 = 10$ is yet another line in the Cartesian plane. (See Fig. 1.2.)

It follows that the common solution or solutions to the equations correspond to points of intersection of the graphs. The system (1.3.29) has a unique solution indicated by the single point of intersection in Figure 1.2. It is a simple matter to find other pairs of equations in two variables that have no common solutions and an infinite number of solutions (see Fig. 1.2). If you think about this simple situation, you quickly see that these are the only possibilities!

1.4 The Gauss-Jordan Elimination Method

The Gauss-Jordan Elimination Method or Algorithm formalizes all of the steps I used to solve the linear systems presented in the last section. Given any nonzero matrix (even if it does not directly correspond to a linear system of equations), the algorithm allows us to compute the *unique* reduced row echelon form (abbreviated **rref**) of a matrix. Here are the steps for finding the rref of A:

1. Use the interchange operation to bring any rows of zeros to the bottom of the matrix.
2. Locate the leftmost nonzero column and focus your attention on the topmost entry (*i.e* the entry in the first row). If this entry is zero, interchange the first row with some

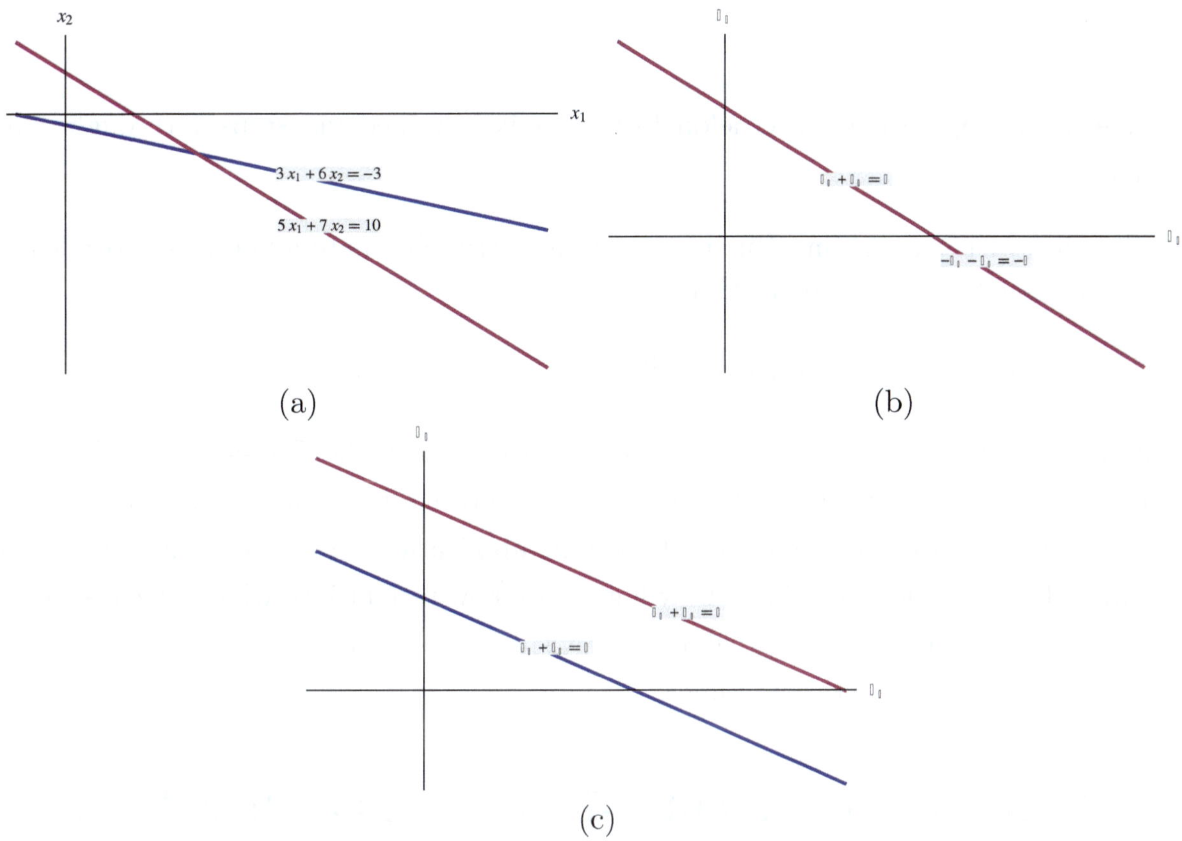

Figure 1.2: Graphs of three possible linear systems with two equations and two unknowns. Points of intersection comprise the solution sets to the equations. In (a), there is a unique solution; in (b), the lines overlap, so the system has an infinite number of solutions; in (c), there are no solutions since we have parallel lines.

other row containing a nonzero leading entry.

3. Use this entry to eliminate all entries below using the elementary row operations of scaling and replacement.

4. Ignore the column and row used in the previous steps. Repeat steps 2 and 3 with the remaining **submatrix**[6] until all rows are exhausted.

This completes the so-called **forward phase** in which we work from the left of A to the right, eliminating entries below leading entries as we go. The result is an eche-

[6]In short, a submatrix is a matrix contained within a larger matrix.

lon form matrix.

The next steps bring the echelon form matrix obtained via steps 1 through 4 into rref:

5. Starting with the rightmost nonzero leading entry; eliminate all entries above leading entries moving to the left of the matrix.

6. Use the scale operation to change all nonzero leading entries to 1's.

I strongly encourage you to revisit our examples to see how the Gauss-Jordan Algorithm is typically used. Before we finish this section, I would like to mention that mastering the Gauss-Jordan Algorithm will be critical to solving applications you'll encounter throughout the course. These applications, in turn, will give you new insight into what it means to solve linear systems of equations. So, we will reuse, revisit, and refine the Gauss-Jordan Algorithm as we continue our journey through linear algebra.

1.5 Application (Calculus-Based): The Heat Equation and Numerical Methods

Many practical applications of mathematics involve modeling real life phenomena with **partial differential equations**. This is a course on linear algebra, so I don't have the time to go into the vast subject of partial differential equations (PDEs). Over the decades, there have been many advances in finding solutions to such equations using paper and pencil. However, a versatile way of attacking a PDE problem modeling a real life situation uses numerical approximation methods.

What does this have to do with linear algebra? Well, it turns out that some PDE problems may be well-approximated by a linear algebra problem! Let's look at an application that can be easily solved using a computer and Gauss-Jordan elimination: suppose we have a square plate of side length l made of some heat-conducting material. We keep the sides of the plate at different temperatures, and we would like to know the **steady-state temperature distribution** of the plate. This is just a fancy way of saying we want to know the temperature at each point on the plate after everything has settled down.

To keep things simple, I'll set the temperature of the left-hand side at $2T_0$ and the temperature of the right side at T_0. The temperature of the top and bottom I'll keep at 0. The

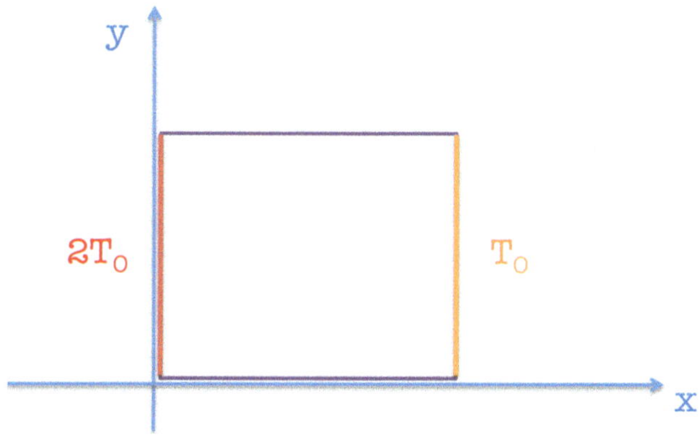

Figure 1.3: Schematic of the heated plate.

temperature distribution of the plate varies with time according to the **heat equation**:

$$\frac{\partial T}{\partial t} = k\left(\frac{\partial^2 T}{\partial x^2} + \frac{\partial^2 T}{\partial y^2}\right) \tag{1.5.1}$$

where k is some constant and T is the temperature, which is a function of space x, y, and time t.

The equation looks intimidating, but don't let the symbols scare you! It is called a partial differential equation, because it relates partial derivatives of a function to each other. The partial derivatives themselves are just like regular derivatives, only you keep all other variables fixed when you take the derivative.

As a quick example, if $f(x, y) = x^2 y + 2y^2$, then

$$\frac{\partial f}{\partial x} = 2xy\ , \ \frac{\partial^2 f}{\partial x^2} = 2y\ , \ \frac{\partial f}{\partial y} = x^2 + 4y. \tag{1.5.2}$$

When everything has settled down, the temperature doesn't change with time, so $\frac{\partial T}{\partial t} = 0$, and we are left with

$$\frac{\partial^2 T}{\partial x^2} + \frac{\partial^2 T}{\partial y^2} = 0. \tag{1.5.3}$$

This is still a very complicated equation to solve! Now we turn this PDE problem into a linear algebra problem. We start by breaking up the plate into a fine grid. For our purposes, I'll just take a 3×3 grid to start with, so you can get a flavor of how this method works.

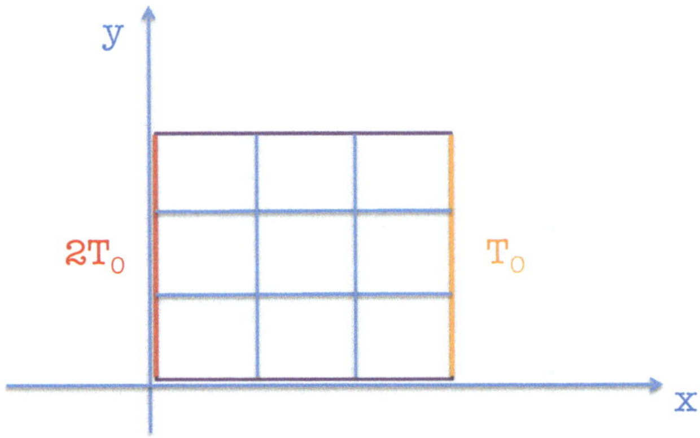

Figure 1.4: Schematic of plate with a 3×3 grid superimposed.

We are going to approximate the derivatives of T at each of the grid points using a basic idea from calculus. Remember that it is possible to approximate a derivative by a **difference quotient**:

$$f'(x) \approx \frac{f(x+h) - f(x)}{h} \tag{1.5.4}$$

The approximation becomes better and better if h is taken smaller and smaller. In a similar way, we can show that the second derivative is well-approximated by

$$f''(x) \approx \frac{f(x+h) - 2f(x) + f(x-h)}{h^2}. \tag{1.5.5}$$

So, if $f(x)$ is defined on some interval $[a, b]$, and we break this interval up into evenly spaced pieces of a one-dimensional grid, we could approximate the second derivative of f at a grid point as follows:

$$f''(\text{grid point}) \approx \frac{f(\text{grid point to the right}) - 2f(\text{grid point}) + f(\text{grid point to the left})}{(\text{length of grid element})^2}. \tag{1.5.6}$$

Notice that this approximation gets better and better the more grid points we add. Another way of looking at it is that the approximation becomes better and better the smaller the size of the grid pieces.

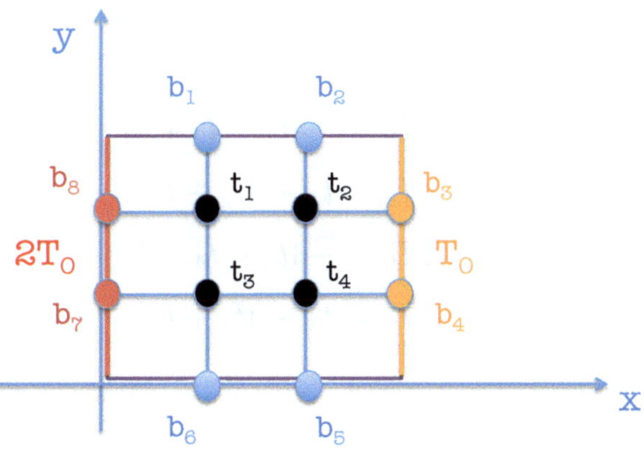

Figure 1.5: Schematic of the heated plate with labels for the temperatures at the grid nodes. Here, b_j represents a boundary temperature, and t_i represents an interior (unknown) temperature.

Now that we know how to approximate second derivatives of functions of single variables, it's a piece of cake to extend this to functions of two variables like T. In fact, we can use this to set up a system of linear equations for the four unknowns t_1, t_2, t_3, t_4 representing the approximate temperature inside the plate! Here are the equations:

$$
\begin{aligned}
0 &= \underbrace{\frac{b_8 - 2t_1 + t_2}{\left(\frac{l}{3}\right)^2}}_{\frac{\partial^2 T}{\partial x^2}} + \underbrace{\frac{b_1 - 2t_1 + t_3}{\left(\frac{l}{3}\right)^2}}_{\frac{\partial^2 T}{\partial y^2}} \\
0 &= \underbrace{\frac{t_1 - 2t_2 + b_3}{\left(\frac{l}{3}\right)^2}}_{\frac{\partial^2 T}{\partial x^2}} + \underbrace{\frac{b_2 - 2t_2 + t_4}{\left(\frac{l}{3}\right)^2}}_{\frac{\partial^2 T}{\partial y^2}} \\
0 &= \underbrace{\frac{b_7 - 2t_3 + t_4}{\left(\frac{l}{3}\right)^2}}_{\frac{\partial^2 T}{\partial x^2}} + \underbrace{\frac{t_1 - 2t_3 + b_6}{\left(\frac{l}{3}\right)^2}}_{\frac{\partial^2 T}{\partial y^2}} \\
0 &= \underbrace{\frac{t_3 - 2t_4 + b_4}{\left(\frac{l}{3}\right)^2}}_{\frac{\partial^2 T}{\partial x^2}} + \underbrace{\frac{t_2 - 2t_4 + b_5}{\left(\frac{l}{3}\right)^2}}_{\frac{\partial^2 T}{\partial y^2}}.
\end{aligned} \tag{1.5.7}
$$

We know the values of the b's. Plugging these values in and clearing the unneeded denominator, we have

$$\begin{aligned}
-2T_0 &= -4t_1 + t_2 + t_3 \\
-T_0 &= t_1 - 4t_2 + t_4 \\
-2T_0 &= -4t_3 + t_4 + t_1 \\
-T_0 &= t_3 - 4t_4 + t_2.
\end{aligned} \tag{1.5.8}$$

This is a linear system with augmented matrix

$$\left(\begin{array}{cccc|c}
-4 & 1 & 1 & 0 & -2T_0 \\
1 & -4 & 0 & 1 & -T_0 \\
1 & 0 & -4 & 1 & -2T_0 \\
0 & 1 & 1 & -4 & -T_0
\end{array}\right). \tag{1.5.9}$$

Applying the Gauss-Jordan algorithm, the augmented matrix has the reduced row echelon form

$$\left(\begin{array}{cccc|c}
1 & 0 & 0 & 0 & \frac{7}{8}T_0 \\
0 & 1 & 0 & 0 & \frac{5}{8}T_0 \\
0 & 0 & 1 & 0 & \frac{7}{8}T_0 \\
0 & 0 & 0 & 1 & \frac{5}{8}T_0
\end{array}\right). \tag{1.5.10}$$

From this, you see that our system has a unique solution. It would be strange indeed if it didn't turn out this way since it would mean that two identical plates with identical temperatures around the edges could have infinitely many different temperature distributions after things settle down. Furthermore, we can read off the temperatures:

$$t_1 = t_3 = \frac{7}{8}T_0 \ , \ t_2 = t_4 = \frac{5}{8}T_0 \tag{1.5.11}$$

Does this make sense? Yes! Since the temperatures are constant along the lengths of the plates, we don't expect the temperature to vary vertically. Also, the grid points closer to the left are hotter than the grid points closer to the right. Also makes sense!

For a more accurate description, we need a more refined grid. A more refined grid means more equations and more unknowns. It gets to the point where a computer could handle the row operations more effectively. That's one reason why computer algorithms for row reducing are so important.

1.6 Chapter 1 Exercises

1. In the introductory section, I presented at a few examples of how linear algebra is used in the real world. We'll encounter more applications as we go along and revisit the ones here to illustrate relevant linear algebra topics. There is a reason you are studying linear algebra. Perhaps you wish to be an engineer, a computer scientist or a financial analyst. Maybe you want to be a mathematician (good for you!). Whatever reason you are studying this wonderful subject, go out and find an application relevant and interesting to you and write a short paragraph about it. Keep it with you for whenever you feel you need extra encouragement along the way to mastering linear algebra!

2. Can you help Alice and Charlie by showing that $x_1 = 14y_1 + 11y_2 \bmod 26$ and $x_2 = 17y_1 + 10y_2 \bmod 26$ give the proper decryption transformation? At the very least, try to convince yourself that this works by decrypting QV and MJ. Remember that $a \bmod N$ (read " a modulo N") is the unique integer between 0 and $N - 1$ obtained by adding or subtracting integer multiples of N to or from a.

3. Prove the general statement that adding a multiple of one equation of a linear system to another does not alter the solution set.

4. In Sally's pizza party problem, show that $a_{11} = 10, a_{12} = 11, a_{13} = 12$, and $b_1 = 65$. Work out the other coefficients and b's for this problem.

5. Examine the graph in the Figure 1.1 and confirm that $1 - 4 - 5$ is the only 2-walk between vertices 1 and 5. Choose a few pairs of vertices from the graph and confirm that the entries of A^2 and A^3 do, in fact, give you the number of 2, and 3-walks, respectively, between the chosen vertices.

6. Use elementary row operations to compute the rref of the following matrices:

(a)

$$\begin{pmatrix} 1 & 3 & -2 \\ -2 & -6 & 4 \\ 6 & -1 & 7 \end{pmatrix}$$

(b)

$$\begin{pmatrix} -2 & 4 & 1 & 3 \\ 3 & -6 & 3 & 0 \\ 0 & 0 & -5 & -5 \end{pmatrix}$$

(c)

$$\begin{pmatrix} 1 & 3 & 5 & 7 & 10 \\ 4 & 8 & 12 & 16 & 0 \\ 5 & 7 & 9 & 1 & 0 \end{pmatrix}$$

(d)

$$\begin{pmatrix} 4 & -8 & -7 & 11 & 2 \\ -2 & 4 & 5 & -5 & 3 \\ 3 & -6 & -6 & 8 & 2 \\ 0 & 0 & 0 & 1 & -1 \end{pmatrix}$$

7. Interpret each matrix of the previous problem as an augmented matrix $(A|\vec{b})$ for a corresponding linear system of equations. In other words, assume the last column corresponds to the b values of a linear system, and the remaining columns give the coefficients for the same set of linear equations. Find all of the solution sets.

8. Using the Gauss-Jordan method on the corresponding augmented matrix, solve the system of equations

$$3x_1 - x_2 + 2x_3 = 1$$
$$2x_1 + x_2 + x_3 = -2$$
$$x_1 - 3x_2 = 3$$

or demonstrate that the system is inconsistent.

9. Using the Gauss-Jordan method on the corresponding augmented matrix, solve the

system of equations

$$
\begin{aligned}
&x_1 + 6x_2 - 18x_3 = 0 \\
&-3x_1 + 6x_2 - 13x_3 = 0 \\
&-4x_1 + 5x_3 = 0 \\
&-7x_1 + 6x_2 - 8x_3 = 1
\end{aligned}
$$

or demonstrate that the system is inconsistent.

10. Assume that each matrix below is the coefficient matrix A for some linear system. Determine all possible $\vec{b}$ so that the corresponding linear system is consistent.

    (a)
    $$\begin{pmatrix} 1 & -2 \\ 0 & -5 \end{pmatrix}$$

    (b)
    $$\begin{pmatrix} 1 & -3 & 0 \\ 3 & -1 & 2 \\ -2 & 6 & 0 \end{pmatrix}$$

    (c)
    $$\begin{pmatrix} 1 & -2 \\ 0 & 6 \\ 4 & 2 \\ 7 & -1 \end{pmatrix}$$

11. Find values for c and d so that the system $x_1 + cx_2 = 7$ and $2x_1 + 4x_2 = d$ has no solutions, a single solution, and infinitely many solutions.

1.7 Chapter 1 References

The reader wanting to know more about linear algebra applications in modern cryptography should consult [1]. The graph theory example was inspired by the book [2]. The heat conduction problem is an example of the more general finite difference method for solving partial differential equations, a topic covered in the book [3]. The medical imaging example is based on the review article [4]. The reader interested in learning more about general applications of linear algebra as well as the Gauss-Jordan algorithm may consult the book [5]. For a more theoretical treatment of the topics covered in this chapter, I suggest consulting [6].

Chapter 2

Vectors and Abstract Vector Spaces

2.1 Column Vectors, Position, Vectors, and Vector Operations

Vectors are quantities with magnitude and direction. This is the simplest way of introducing the idea of a vector. You experience vector quantities everyday: forces like gravity pulling you down to the earth, the wind and water currents you feel at the beach, even the light entering your eye from a sunrise can all be represented by vectors.

All of these quantities have magnitude (or strength) and direction: if you lose a few pounds, the force gravity on you lessens, but still pulls directly downwards. "But wait!" you might say. "Didn't you define a vector as a matrix with a single column back when Alice and Charlie came up with their encryption scheme?" You are quite right; I did define a **column vector** as matrix of size $n \times 1$. So, is it a coincidence or bad terminology that mathematicians use the same name? Or, is there an actual connection between the two ideas? It turns out there is a connection: it is possible to geometrically interpret an $n \times 1$ column vector as an arrow in $\mathbb{R}^n$ that has both magnitude (length) and direction.

It's easiest to think about this in two dimensions. Remember that the Cartesian plane (or $\mathbb{R}^2$ if you like) consists of points with coordinates given by ordered pairs of numbers (x_1, x_2). Here, the parentheses are meant to indicate an ordered list. However, the same information is contained in the column vector

$$\begin{pmatrix} x_1 \\ x_2 \end{pmatrix}, \tag{2.1.1}$$

so we have an equivalency

$$\begin{pmatrix} x_1 \\ x_2 \end{pmatrix} \leftrightarrow (x_1, x_2), \tag{2.1.2}$$

where, on the one hand, we have a geometric point in the plane and on the other, we have a column vector of size 2×1. To each point $P = (x_1, x_2)$ in the plane, it is possible to assign an arrow with a tail at the origin and a tip at the point P. Let's denote this arrow by $\vec{P}$. (See Fig. 2.1.)

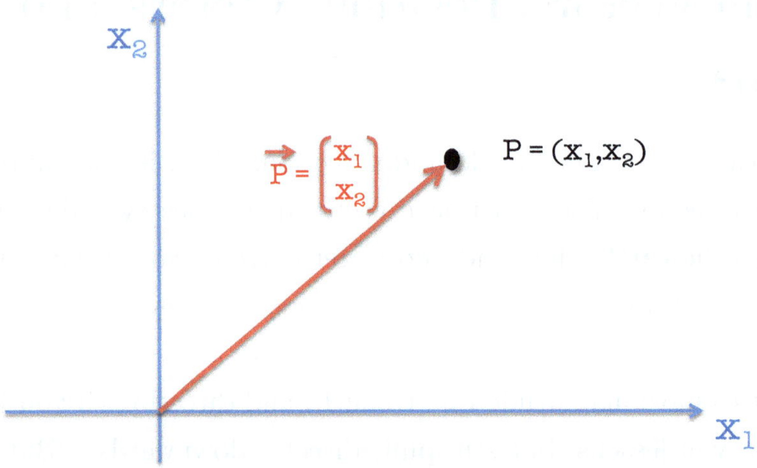

Figure 2.1: A point $P = (x_1, x_2)$ in the Cartesian plane may be considered equivalent to a position vector $\vec{P} \in \mathbb{R}^2$.

Since points are equivalent to column vectors, the arrow also represents the column vector $\begin{pmatrix} x_1 \\ x_2 \end{pmatrix}$, so we write $\vec{P} = \begin{pmatrix} x_1 \\ x_2 \end{pmatrix}$ for the **position vector of P**. For the first part of this chapter, vectors will have the dual interpretation of the geometric arrow and the column vector, so I'll retain the overhead arrow notation for such vectors. When we come to abstract vectors, I'll switch to a slightly different notation.

Even though we cannot visualize higher dimensions, we can still generalize the above discussion to $\mathbb{R}^n$. In other words, a point P in $\mathbb{R}^n$ represented by the n-tuple of numbers $(x_1, \ldots, x_n)$ is equivalent to the column vector

$$\vec{P} = \begin{pmatrix} x_1 \\ x_2 \\ \vdots \\ x_n \end{pmatrix}, \tag{2.1.3}$$

which, in turn, can be thought of as an arrow from the origin in $\mathbb{R}^n$ to the point P.

Thinking of vectors as arrows can be very useful in solving many linear algebra problems. Before we can apply vectors to such problems, I need to establish some rules, or **operations**, on vectors. First, it is possible to define a sensible operation of **vector addition** in the following way: I can write any two vectors $\vec{u}, \vec{v} \in \mathbb{R}^n$ in terms of their components:

$$\vec{u} = \begin{pmatrix} u_1 \\ u_2 \\ \vdots \\ u_n \end{pmatrix}, \vec{v} = \begin{pmatrix} v_1 \\ v_2 \\ \vdots \\ v_n \end{pmatrix}. \tag{2.1.4}$$

It is then quite natural to define a new vector called $\vec{u} + \vec{v}$ given by

$$\vec{u} + \vec{v} = \begin{pmatrix} u_1 + v_1 \\ u_2 + v_2 \\ \vdots \\ u_n + v_n \end{pmatrix}; \tag{2.1.5}$$

to add $\vec{u}$ and $\vec{v}$, you simply add corresponding components and form the column vector from the resulting additions. Notice that the definition only makes sense if the vectors have the same number of entries; i.e. live in the same $\mathbb{R}^n$.

Example: If $\vec{u} = \begin{pmatrix} -1 \\ 3 \\ 0 \end{pmatrix}, \vec{v} = \begin{pmatrix} 2 \\ 5 \\ -1 \end{pmatrix} \in \mathbb{R}^3$, then $\vec{u} + \vec{v} = \begin{pmatrix} 1 \\ 8 \\ -1 \end{pmatrix}$.

The **parallelogram rule** allows us to visualize the process of vector addition, at least in $\mathbb{R}^2$ and $\mathbb{R}^3$. Let's focus on $\mathbb{R}^2$, since it is easiest to draw and visualize. The first step in applying the parallelogram rule is to draw the arrows representing $\vec{u}$ and $\vec{v}$ in the plane as in Fig. 2.2. Then, a copy of $\vec{u}$ (the red vector in Fig. 2.2) [1] is made and moved without altering it or rotating it so that the base sits at the tip $\vec{v}$. The arrow from the origin to the tip of the copied and translated $\vec{u}$ (in red) represents $\vec{u} + \vec{v}$ (in green). This is the parallelogram rule for vector addition. Another way of thinking about this rule (and also explaining the name) is to suppose that the arrows $\vec{u}$ and $\vec{v}$ define a parallelogram in the plane in which $\vec{u} + \vec{v}$ represents the diagonal of the parallelogram coming from the shared vertex of $\vec{u}$ and $\vec{v}$.

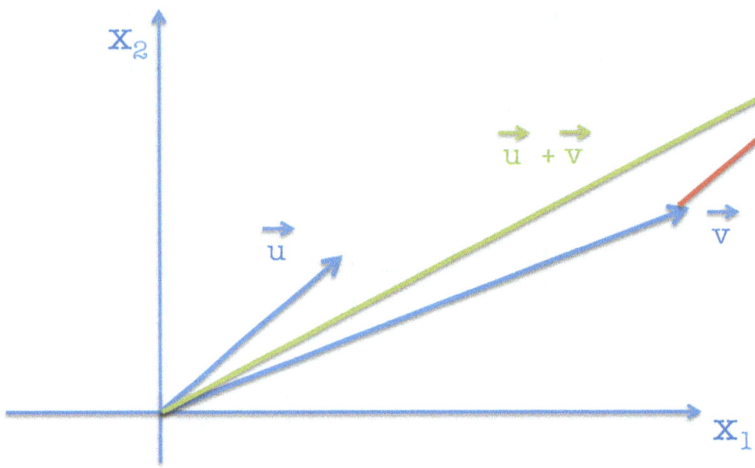

Figure 2.2: The Parallelogram Rule: the sum of two vectors in the plane $\vec{u}$ and $\vec{v}$ can be represented geometrically by copying $\vec{u}$ and translating it without alteration so that the base lines up with the tip of $\vec{v}$. The copied vector is depicted here in red. The vector sum $\vec{u}+\vec{v}$ (in green) is the arrow that originates at the origin and stops at the tip of the copied/translated $\vec{u}$.

Along with the operation of vector addition, we also have the **operation of scalar multiplication**. Basically, this allows us to define $c\vec{v}$ whenever $c \in \mathbb{R}$. Just like vector addition,

[1] It doesn't matter if we choose $\vec{u}$ or $\vec{v}$ for this process, but I'll choose $\vec{u}$ for concreteness.

scalar multiplication has a very nice and intuitive definition:

$$c\vec{u} = \begin{pmatrix} cu_1 \\ cu_2 \\ \vdots \\ cu_n \end{pmatrix}; \tag{2.1.6}$$

i.e. to compute $c\vec{u}$, you simply multiply every component by c.

Example: Given $\vec{u} = \begin{pmatrix} -1 \\ 3 \\ 0 \end{pmatrix}$, it follows that $2\vec{u} = \begin{pmatrix} -2 \\ 6 \\ 0 \end{pmatrix}$ and $-\frac{1}{2}\vec{u} = \begin{pmatrix} \frac{1}{2} \\ -\frac{3}{2} \\ 0 \end{pmatrix}$.

Geometrically, scalar multiplication is easier to interpret than vector addition. I will now present at a few separate cases. First, if $c > 1$, then the overall effect of $c\vec{u}$ is to stretch $\vec{u}$ by a factor of c. For instance, if $c = 2$, then $2\vec{u}$ is a vector that points in the same direction as $\vec{u}$, but has double the length[2]. On the other hand, if $0 < c < 1$, then the effect of $c\vec{u}$ is to shrink the vector; $\frac{1}{2}\vec{u}$ points in the same direction as $\vec{u}$ but only has half the length. (See Fig. 2.3.)

When $c < 0$, we can always write $c = (-1)|c|$ where $|c| > 0$. In this case, we can think of $c\vec{u}$ as being made up of two scalar multiplications; $|c|\vec{u}$ and then $(-1)|c|\vec{u}$. $|c|\vec{u}$ leaves the direction the same, but changes the overall length. The effect of multiplying by -1 is to leave the *length* alone but *reverse* the direction.

Now that I have defined scalar multiplication and vector addition, you know how to compute $a\vec{u} + b\vec{v}$ for any given $a, b \in \mathbb{R}$ and any $\vec{u}, \vec{v} \in \mathbb{R}^n$. This vector is known as a **linear combination** of $\vec{u}, \vec{v}$ with **weights** a and b, respectively.

Definition: Given a set of vectors $\{\vec{u}_1, \ldots, \vec{u}_p\}$ all belonging to $\mathbb{R}^n$, the vector

$$\vec{x} = c_1\vec{u}_1 + \cdots + c_p\vec{u}_p \tag{2.1.7}$$

[2]Technically, I haven't defined the length of a vector yet, but I will trust your intuitive notion of vector length for the moment.

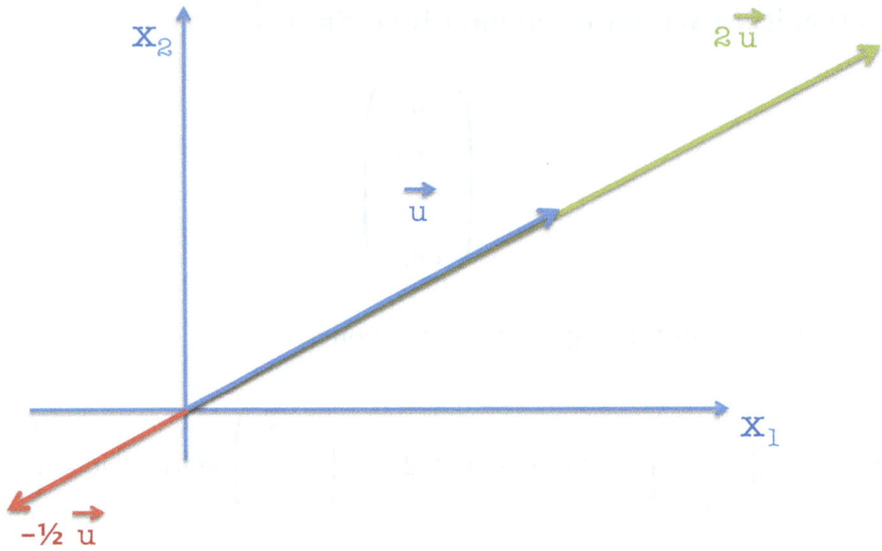

Figure 2.3: Geometric depictions of a nonzero vector $\vec{u}$ (blue) multiplied by a factor of 2 (green) and $-1/2$ (red).

is called the **linear combination** of $\{\vec{u}_1, \ldots, \vec{u}_p\}$ with **weights** $c_1, \ldots, c_p$.

In a sense, the weights in a linear combination determine how much each vector contributes to the whole; if say $c_1 = 5$ and $c_2 = c_3 = 0.001$ and $\vec{u}_1, \vec{u}_2$ and $\vec{u}_3$ all have unit length, it is not hard to convince yourself that $c_1\vec{u}_1 + c_2\vec{u}_2 + c_3\vec{u}_3$ will be a vector that mostly points in the direction of $\vec{u}_1$ and has a length pretty close to 5.[3] Right now, there are no constraints on the possible values of the weights (other than the assumption that they are real numbers). If we let the weights take on all possible values, we have an infinite family of vectors called the **span** of $\{\vec{u}_1, \ldots, \vec{u}_p\}$:

Definition: The **span** of $\{\vec{u}_1, \ldots, \vec{u}_p\}$, denoted by $\text{Span}\{\vec{u}_1, \ldots, \vec{u}_p\}$, is defined to be the set of all linear combinations of $\{\vec{u}_1, \ldots, \vec{u}_p\}$:

$$\text{Span}\{\vec{u}_1, \ldots, \vec{u}_p\} = \{\vec{x} \in \mathbb{R}^n | \vec{x} = c_1\vec{u}_1 + \cdots + c_p\vec{u}_p, c_j \in \mathbb{R}\}. \tag{2.1.8}$$

[3]You can also think of this idea in terms of exam averages: if the final exam is worth much more than the other exams during the semester, your performance on the final impacts your overall grade with greatest weight.

The definition of the span of a set of vectors is easy to state but challenging to interpret. I'll present a few basic examples to give you a sense of what the term "span" actually means.

Example: Span of a Single Vector

To keep things as simple as possible, consider a nonzero vector $\vec{u}$ in $\mathbb{R}^2$. By definition,

$$\text{Span}\{\vec{u}\} = \{c\vec{u} \mid c \in \mathbb{R}\}, \tag{2.1.9}$$

which is just the set of all scalar multiples of $\vec{u}$. As we know, $c\vec{u}$ is parallel or anti-parallel to $\vec{u}$ (depending on the sign of c), and the magnitude of c determines if the vector is shrunk or stretched. If you imagine all of these vectors drawn out in $\mathbb{R}^2$, you get a line parallel to $\vec{u}$ and through the origin. (See Fig. 2.4.)

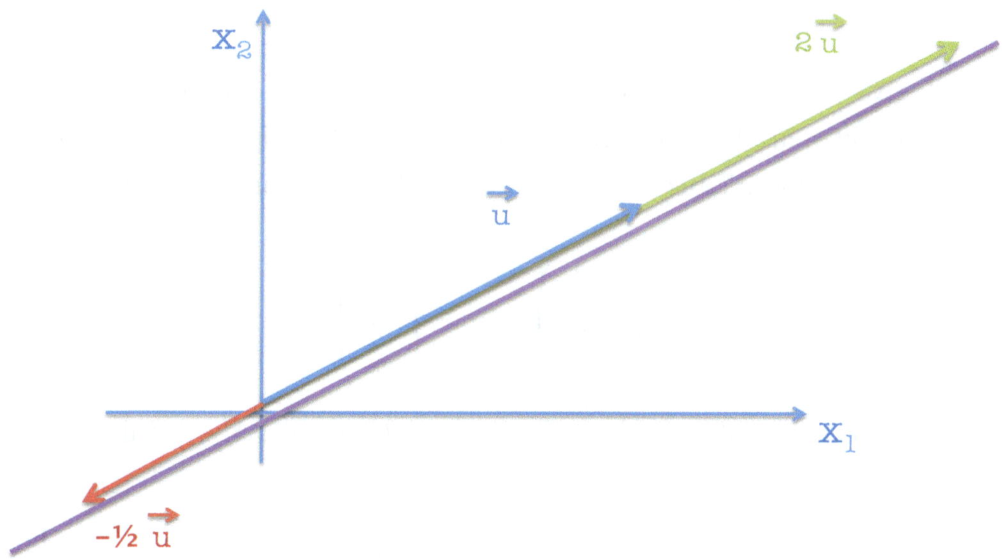

Figure 2.4: All scalar multiples of a single nonzero vector $\vec{u}$ in $\mathbb{R}^2$ together define a line through the origin (purple) parallel to $\vec{u}$.

The same conclusion holds if $\vec{u} \in \mathbb{R}^n$, namely the span of $\vec{u}$ is a line parallel to $\vec{u}$ through the origin. In general, if you wish to find the line passing through two distinct points $\vec{P}$ and $\vec{Q}$, you may use the general result:

Theorem: A point $\vec{x}$ is on the line through $\vec{P}$ and $\vec{Q}$ if and only if there exists some $t \in \mathbb{R}$ such that

$$\vec{x} = t\left(\vec{P} - \vec{Q}\right) + \vec{Q}. \tag{2.1.10}$$

The basic idea behind this theorem is that the line going through $\vec{P}$ and $\vec{Q}$ runs along the **displacement vector** $\vec{P} - \vec{Q}$. By stretching, shrinking, and reversing $\vec{P} - \vec{Q}$ you can reach every point on the line. In the case where the line is described by $\text{Span}\{\vec{u}\}$ with $\vec{u} \neq \vec{0}$, you can choose $\vec{P} = \vec{u}$ and $\vec{Q} = \vec{0}$. Suppose you need to multiply $\vec{P} - \vec{Q}$ by t_0 to reach x, a point on the line. By the parallelogram rule, the vector $\vec{x}$ representing point x on the line is given by $\vec{x} = t_0\left(\vec{P} - \vec{Q}\right) + \vec{Q}$. Conversely, every vector of the form $t\left(\vec{P} - \vec{Q}\right) + \vec{Q}$ with $t \in \mathbb{R}$ represents a point on the line.

Example: Span of Two Vectors

Suppose that I give you two nonzero vectors $\vec{u}, \vec{v} \in \mathbb{R}^3$, and suppose further that they are not parallel or anti-parallel. In other words, assume there is no $t \in \mathbb{R}$ where $\vec{u} = t\vec{v}$ is true. So the discussion doesn't become too abstract, I'll pick concrete column vectors for $\vec{u}$ and $\vec{v}$, say

$$\vec{u} = \begin{pmatrix} -1 \\ 3 \\ 0 \end{pmatrix}, \quad \vec{v} = \begin{pmatrix} 2 \\ 0 \\ 1 \end{pmatrix}. \tag{2.1.11}$$

It's easy to show that there is no $t \in \mathbb{R}$ such that $\vec{u} = t\vec{v}$ is true. By definition,

$$\text{Span}\{\vec{u}, \vec{v}\} = \left\{ t\begin{pmatrix} -1 \\ 3 \\ 0 \end{pmatrix} + s\begin{pmatrix} 2 \\ 0 \\ 1 \end{pmatrix} \middle| t, s \in \mathbb{R} \right\} = \left\{ \begin{pmatrix} -t + 2s \\ 3t \\ s \end{pmatrix} \middle| t, s \in \mathbb{R} \right\} \tag{2.1.12}$$

where I have chosen the weights to be represented by t and s, respectively, to make things as clear as possible. The $\text{Span}\{\vec{u}, \vec{v}\}$ therefore consists of vectors where $x_1 = -t + 2s$, $x_2 = 3t$ and $x_3 = s$ for some choice of t and s. Here, the weights t and s are viewed as parameters. We can eliminate the parameters entirely by back-substitution if we wish:

$$x_1 = -\frac{1}{3}x_2 + 2x_3 \tag{2.1.13}$$

or

$$x_1 + \frac{1}{3}x_2 - 2x_3 = 0. \tag{2.1.14}$$

If you find all possible points (x_1, x_2, x_3) with coordinates satisfying $x_1 + \frac{1}{3}x_2 - 2x_3 = 0$ and plot these points in $\mathbb{R}^3$, you construct a **plane** passing through the origin. Geometrically, you can think of the plane as an infinitely wide and long flat sheet of paper tilted with respect to the coordinate axes in $\mathbb{R}^3$ and slicing through the origin. (See Fig. 2.5.)

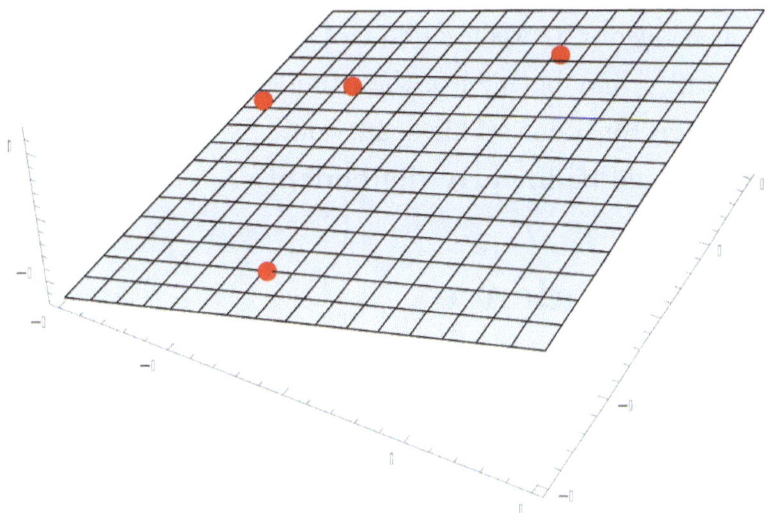

Figure 2.5: If we select a few x_1 and x_2 values at random, we can generate a few points in the set Span$\{\vec{u}, \vec{v}\}$ through the equation $x_1 + \frac{1}{3}x_2 - 2x_3 = 0$. These points are depicted as red dots. Doing this for all possible values of x_1 and x_2 generates a plane through the origin, part of which is depicted here as a blue "sheet" cutting through the origin.

In both of the examples, you see that spanning sets represent geometric objects that, in a sense, don't bend. In the first case where we had the spanning set of a single nonzero vector, the span turned out to be a line. In the second case, with two nonparallel spanning vectors, the span was a plane. As geometric objects, the line and the plane don't have bumps, hills, or folds. In higher dimensions with larger spanning sets, the same idea carries over, but it is much more difficult to visualize. I'll have more to say about the geometric interpretation of spanning sets when we come to **vector subspaces**.

In order for our spanning set Span$\{\vec{u}, \vec{v}\}$ to represent a plane, I chose nonzero vectors $\vec{u}$

and $\vec{v}$ which were nonparallel. In other words, there is no real number $t \in \mathbb{R}$ where $\vec{u} = t\vec{v}$ is true. Another way of expressing this idea is to say that there are no solutions to

$$c_1\vec{u} + c_2\vec{v} = 0 \tag{2.1.15}$$

other than the **trivial solution** $c_1 = c_2 = 0$. In other words, $\vec{u}$ and $\vec{v}$ were selected to be **linearly independent**.

If there were a linear relationship between $\vec{u}$ and $\vec{v}$ of the form $\vec{u} = t\vec{v}$, $t \neq 0$ or (equivalently) $c_1\vec{u} + c_2\vec{v} = \vec{0}$ with c_1 and c_2 not both zero, then the span of the two vectors would be the same as the span of one vector alone. This span would be a *line* in $\mathbb{R}^3$ through the origin. In this case, the vectors would be **linearly dependent**.

2.2 Linear Independence

The linear independence idea seen in the plane example of the last section generalizes to arbitrary sets of vectors:

Definition: A set of vectors $\{\vec{u}_1, \dots, \vec{u}_p\}$ in $\mathbb{R}^n$ is said to be **linearly independent** if the only solution to the vector equation

$$c_1\vec{u}_1 + \cdots + c_p\vec{u}_p = \vec{0} \tag{2.2.1}$$

is the **trivial solution** $c_1 = c_2 = \cdots = c_p = 0$. Here, $\vec{0}$ is the vector in $\mathbb{R}^n$ with entries all 0. Otherwise, the set of vectors is said to be **linearly dependent**.

Before I explore the geometric significance of linear independence, I'll spend a little time developing a technique for determining the linear independence of a set of vectors. To help you along, I'll look at a specific example. Take

$$\vec{u} = \begin{pmatrix} -1 \\ 3 \\ 0 \end{pmatrix}, \ \vec{v} = \begin{pmatrix} 2 \\ 0 \\ 1 \end{pmatrix} \tag{2.2.2}$$

as before.

The vector equation $c_1\vec{u} + c_2\vec{v} = \vec{0}$ leads to a system of linear equations in c_1 and c_2:

$$c_1\vec{u} + c_2\vec{v} = \vec{0} \to c_1 \begin{pmatrix} -1 \\ 3 \\ 0 \end{pmatrix} + c_2 \begin{pmatrix} 2 \\ 0 \\ 1 \end{pmatrix} = \begin{pmatrix} 0 \\ 0 \\ 0 \end{pmatrix} \tag{2.2.3}$$

or

$$\begin{pmatrix} -c_1 + 2c_2 \\ 3c_1 \\ c_2 \end{pmatrix} = \begin{pmatrix} 0 \\ 0 \\ 0 \end{pmatrix}. \tag{2.2.4}$$

By comparing the components on the left and on the right, we have

$$\begin{aligned} -c_1 + 2c_2 &= 0 \\ 3c_1 &= 0 \\ c_2 &= 0. \end{aligned} \tag{2.2.5}$$

It's not too hard to stare at these equations and realize that the only answer is $c_1 = c_2 = 0$, establishing the linear independence of $\vec{u}$ and $\vec{v}$. But, we could equally well solve this system by using the Gauss-Jordan algorithm. Setting up the augmented matrix

$$\left(\begin{array}{cc|c} -1 & 2 & 0 \\ 3 & 0 & 0 \\ 0 & 1 & 0 \end{array}\right), \tag{2.2.6}$$

the row reduced echelon form is

$$\left(\begin{array}{cc|c} 1 & 0 & 0 \\ 0 & 1 & 0 \\ 0 & 0 & 0 \end{array}\right). \tag{2.2.7}$$

The only solution is $c_1 = c_2 = 0$.

Notice something very important: to establish the linear independence or dependence of $\vec{u}$ and $\vec{v}$, I built the augmented matrix $\left(\vec{u}, \vec{v} \mid \vec{0}\right)$ and performed elementary row operations to find the unique reduced row echelon form. In other words, I took the vectors under consideration and put them as columns in an augmented matrix with the last column being a column of zeros.

Suppose, instead, that we have a collection of vectors $\{\vec{u}_1, \ldots, \vec{u}_p\}$, and we wish to know if they are linearly independent or dependent. From the above argument, we have the following theorem:

Theorem: Given a collection of vectors $\{\vec{u}_1, \ldots, \vec{u}_p\}$ in $\mathbb{R}^n$, form the augmented matrix

$$\left(\vec{u}_1, \ldots, \vec{u}_p \mid \vec{0}\right). \tag{2.2.8}$$

The vectors are linearly independent if and only if this linear system has a unique solution, namely the trivial solution $\vec{0}$. Otherwise, the vectors are linearly dependent.

I'll take this opportunity to introduce some terminology now that will benefit us later.

Definition: A linear system represented by the augmented matrix $(A \mid \vec{0})$ where A is the coefficient matrix is called a **homogeneous system**.

You see that homogeneous systems naturally arise if we are trying to understand the linear dependence or independence of a collection of vectors. One of the great features about homogeneous systems is that they are always consistent: the trivial solution consisting of all zeros is *always* a solution. No matter what the coefficients are in A, if you plug in 0 for all variables, the result is 0 for each equation.

The real question here is whether the homogeneous system has solutions other than the trivial one. If so, then the system possesses an infinite number of so-called **nontrivial solutions**, i.e. solutions where not all of the variables are 0.

Example: Determine if

$$\vec{u}_1 = \begin{pmatrix} 1 \\ -1 \end{pmatrix}, \quad \vec{u}_2 = \begin{pmatrix} 2 \\ 3 \end{pmatrix}, \quad \vec{u}_3 = \begin{pmatrix} 6 \\ 4 \end{pmatrix} \tag{2.2.9}$$

are linearly independent or dependent. If they are dependent, find a linear relationship between the vectors.

Start by constructing the augmented matrix

$$\left(\begin{array}{ccc|c} 1 & 2 & 6 & 0 \\ -1 & 3 & 4 & 0 \end{array}\right). \tag{2.2.10}$$

According to the Gauss-Jordan algorithm, I can use the 1 in the first row to eliminate the -1 below it:

$$\left(\begin{array}{ccc|c} 1 & 2 & 6 & 0 \\ -1 & 3 & 4 & 0 \end{array}\right) \begin{array}{c} r_1 \to r_1 \\ r_2 \to r_2 + r_1 \end{array} \left(\begin{array}{ccc|c} 1 & 2 & 6 & 0 \\ 0 & 5 & 10 & 0 \end{array}\right). \tag{2.2.11}$$

I'll scale the second row and use row 2 to complete the backward phase of the Gauss-Jordan algorithm:

$$\left(\begin{array}{ccc|c} 1 & 2 & 6 & 0 \\ 0 & 5 & 10 & 0 \end{array}\right) \begin{array}{c} r_2 \to \left(\frac{1}{5}\right) r_2 \\ r_1 \to r_1 - 2r_2 \end{array} \left(\begin{array}{ccc|c} 1 & 0 & 2 & 0 \\ 0 & 1 & 2 & 0 \end{array}\right). \tag{2.2.12}$$

From the rref, we see that the solution set consists of an infinite number of elements: the equivalent linear system tells us that

$$\begin{aligned} c_1 + 2c_3 &= 0 \\ c_2 + 2c_3 &= 0 \end{aligned} \tag{2.2.13}$$

with no constraint on the value of c_3: we can choose any (nonzero) value we wish for c_3 and, as long as $c_1 = -2c_3$ and $c_2 = -2c_3$, we have a solution to the original linear system. To keep things simple, I'll take $c_3 = 1$, which means that $c_1 = c_2 = -2$. Then, I claim that

$$(-2 = c_1) \begin{pmatrix} 1 \\ -1 \end{pmatrix} + (-2 = c_2) \begin{pmatrix} 2 \\ 3 \end{pmatrix} + (1 = c_3) \begin{pmatrix} 6 \\ 4 \end{pmatrix} = \begin{pmatrix} 0 \\ 0 \end{pmatrix}, \tag{2.2.14}$$

which you can easily verify is true.

This is one possible linear relationship between the vectors among an infinite number. There will be a linear relationship for each nonzero choice of c_3. Try constructing a few others, and verify that they are true.

I finish this discussion on linear independence with a theorem we'll find useful later:

Theorem: If a set of vectors $\{\vec{u}_1, \ldots, \vec{u}_p\}$ in $\mathbb{R}^n$ has more vectors than components (i.e. $p > n$), then the set of vectors is automatically linearly dependent.

The proof of this statement is fairly simple to describe. The vector equation $c_1\vec{u}_1 + \cdots + c_p\vec{u}_p = \vec{0}$ leads to a consistent homogeneous system $\left(\vec{u}_1, \ldots, \vec{u}_p \mid \vec{0}\right)$. This matrix is $n \times (p+1)$ where $p > n$. If we perform the Gauss-Jordan algorithm on this matrix to find the unique rref, then we can expect at most n leading ones; i.e. we cannot have more leading entries than there are rows. This means that at least one of the variables (i.e. the weights) c_j is free to take on any real value, leading to an infinite number of solutions and, hence, at least one nontrivial solution. A nontrivial solution means the original set of vectors is necessarily linearly dependent.

Now that we have a pretty clear idea of how to determine if a set of vectors is linearly independent and what that means from a conceptual standpoint, I'll present a very important set of ideas that follow directly from linear independence: basis vectors and coordinate vectors.

Basis Vectors and Coordinate Vectors

Imagine that Bill lives in a very ordered city where the streets are laid out in a nice grid. We can think of Bill living in $\mathbb{R}^2$ if we assume he can only move in the x_1- and x_2-directions (i.e. he doesn't have a jet pack that allows him to fly up and down). Bill navigates through his $\mathbb{R}^2$ city by using the coordinates (x_1, x_2) based on some fixed set of coordinate axes with the origin, $(0, 0)$, at Bill's house. Among Bill's favorite destinations is a restaurant at $(1, 2)$. Assuming the streets are laid out on a grid, this means Bill has to travel 1 unit up the x_1-direction and 2 units up the x_2-direction to get to the restaurant.

Given what we know about the relationship between points and vectors, we can think about Bill's favorite restaurant in a slightly different but useful way. If we think of $\vec{S}$ as representing the position of the restaurant, then we can write (using the operations of vector addition and scalar multiplication)

$$\vec{S} = \begin{pmatrix} 1 \\ 2 \end{pmatrix} = 1 \begin{pmatrix} 1 \\ 0 \end{pmatrix} + 2 \begin{pmatrix} 0 \\ 1 \end{pmatrix} = \hat{e}_1 + 2\hat{e}_2. \tag{2.2.15}$$

Here, I have defined $\hat{e}_i$ to be vector with unit length running along the positive x_i-direction. The "hat" on top of the vector is supposed to indicate that it has unit length. In general, $\hat{e}_j \in \mathbb{R}^n$ is called the j-th **standard basis vector**. Again, the hat means that the arrow has unit length. It is a vector made up of zeros except for a single entry of 1: the subscript j tells you where this one should appear.

Example: For $\mathbb{R}^3$, we have

$$\hat{e}_1 = \begin{pmatrix} 1 \\ 0 \\ 0 \end{pmatrix}, \ \hat{e}_2 = \begin{pmatrix} 0 \\ 1 \\ 0 \end{pmatrix}, \ \hat{e}_3 = \begin{pmatrix} 0 \\ 0 \\ 1 \end{pmatrix}, \tag{2.2.16}$$

and for $\mathbb{R}^4$, we have

$$\hat{e}_1 = \begin{pmatrix} 1 \\ 0 \\ 0 \\ 0 \end{pmatrix}, \ \hat{e}_2 = \begin{pmatrix} 0 \\ 1 \\ 0 \\ 0 \end{pmatrix}, \ \hat{e}_3 = \begin{pmatrix} 0 \\ 0 \\ 1 \\ 0 \end{pmatrix}, \ \hat{e}_4 = \begin{pmatrix} 0 \\ 0 \\ 0 \\ 1 \end{pmatrix}. \tag{2.2.17}$$

From $\vec{S}$, we get a sense of why $\hat{e}_1$ and $\hat{e}_2$ are called basis vectors for $\mathbb{R}^2$: we can build from them every other vector in $\mathbb{R}^2$ using the operations of scalar multiplication and vector addition. To get the vector $\vec{S}$, we simply stretch $\hat{e}_1$ by a factor of 1 (which leaves the vector $\hat{e}_1$ as it is) and then stretch the vector $\hat{e}_2$ by a factor of 2 and then add the result.

For $\mathbb{R}^2$, we can write

$$\vec{x} = \begin{pmatrix} x_1 \\ x_2 \end{pmatrix} = x_1 \begin{pmatrix} 1 \\ 0 \end{pmatrix} + x_2 \begin{pmatrix} 0 \\ 1 \end{pmatrix} = x_1\hat{e}_1 + x_2\hat{e}_2 \tag{2.2.18}$$

and, for $\mathbb{R}^n$, we can write

$$\vec{x} = x_1\hat{e}_1 + \cdots + x_n\hat{e}_n \tag{2.2.19}$$

as the reader should verify.

Building on our fictional example, imagine Bill's restaurant occupies the top floor of a building 3 units high. Then, the precise coordinates of the restaurant in $\mathbb{R}^3$ would be $(1, 2, 3)$ corresponding to the position vector $\vec{S} = \hat{e}_1 + 2\hat{e}_2 + 3\hat{e}_3$. Hopefully you see that the coordinates of a point S in the chosen $x_1 - x_2 - x_3$ coordinate system can be thought of as the weights for the standard basis vectors describing the corresponding position vector $\vec{S}$.

You may naturally wonder if the original $x_1 - x_2$ coordinate system in which the coordinate axes make right angles with respect to each other is the only possible choice of coordinate system. In other words, did I have to use the $x_1 - x_2$ coordinate system to describe Bill's city? The answer is no, the coordinate system was my choice, since Bill's city is laid out on a grid, and I found this coordinate system *most useful.* If Bill's hometown were laid out so that the cross streets met at a 45° angle, then it would make sense to use a coordinate system where the axes also met at a 45° angle.

This problem of finding or choosing a convenient coordinate system comes up in many different situations. The general goal is to find some way of assigning a *unique* finite list of real numbers to every object in a mathematical problem in much the same way that the restaurant's location in the grid was described as $(1, 2)$. I emphasize unique, since having two or more lists of numbers describing your object is not useful. If you were living in Bill's city, and you wanted directions to this restaurant he's always raving about, you would scratch your head if he were to say "It is located at $(1, 2)$ and at $(3, -1)$". A single and unambiguous list of numbers is the best possible description for our purposes.

Using the vectors $\hat{e}_1$ and $\hat{e}_2$ in $\mathbb{R}^2$ gives us a single unambiguous list of coordinates for any point in $\mathbb{R}^2$ for two important reasons: first, *any* vector can be written as at least one linear combination of $\hat{e}_1$ and $\hat{e}_2$ (giving us at least one set of coordinates by using the weights) and, second, any linear combination in terms of $\hat{e}_1$ and $\hat{e}_2$ is *unique.* These are the necessary ingredients for a **basis**.

Definition: A **basis** for $\mathbb{R}^n$ is a set of linearly independent vectors $\{\vec{u}_1, \ldots, \vec{u}_n\}$ such that $\mathbb{R}^n = \text{Span}\{\vec{u}_1, \ldots, \vec{u}_n\}$.

Notice that we need n basis vectors to describe the set of all n-tuples of real numbers $\mathbb{R}^n$. I'll show you why this has to be the case soon. For now, just accept it as part of the definition[4].

Given $\vec{x} \in \mathbb{R}^n$ (which we think of as a geometric arrow in $\mathbb{R}^n$), this vector has a unique description in terms of **coordinates with respect to the basis** $\mathcal{B} = \{\vec{u}_1, \ldots, \vec{u}_n\}$:

Definition: The **coordinate vector of $\vec{x}$ with respect to the basis $\mathcal{B}$** $= \{\vec{u}_1, \ldots, \vec{u}_n\}$ is defined to be

$$[\vec{x}]_{\mathcal{B}} = \begin{pmatrix} c_1 \\ \vdots \\ c_n \end{pmatrix} \tag{2.2.20}$$

where

$$\vec{x} = c_1\vec{u}_1 + \cdots + c_n\vec{u}_n \tag{2.2.21}$$

is the unique description of $\vec{x}$ in terms of the basis vectors in $\mathcal{B}$.

Example: Consider the set $\mathcal{B} = \left\{ \begin{pmatrix} 1 \\ 1 \end{pmatrix}, \begin{pmatrix} 1 \\ -1 \end{pmatrix} \right\}$. This collection of vectors is linearly independent and forms a basis for $\mathbb{R}^2$. It is not too hard to show that $\vec{x} = \begin{pmatrix} 1 \\ 2 \end{pmatrix}$ can be uniquely written as

$$\begin{pmatrix} 1 \\ 2 \end{pmatrix} = \frac{3}{2}\begin{pmatrix} 1 \\ 1 \end{pmatrix} + \left(-\frac{1}{2}\right)\begin{pmatrix} 1 \\ -1 \end{pmatrix}, \tag{2.2.22}$$

so

$$[\vec{x}]_{\mathcal{B}} = \begin{pmatrix} \frac{3}{2} \\ -\frac{1}{2} \end{pmatrix}. \tag{2.2.23}$$

[4]Give some thought as to why more or fewer vectors do not give unambiguous coordinates, so, do not form a useful basis.

It can be confusing to write $\vec{x}$ as $\begin{pmatrix} \frac{3}{2} \\ -\frac{1}{2} \end{pmatrix}$ when we defined it to be $\begin{pmatrix} 1 \\ 2 \end{pmatrix}$ originally. It helps to keep in mind that vectors are geometric objects that exist independently of the way we describe them (we've been drawing them as arrows, after all). In this case $\vec{x}$ may be described as $\begin{pmatrix} \frac{3}{2} \\ -\frac{1}{2} \end{pmatrix}$ if the basis $\mathcal{B}$ is convenient to use or as $\begin{pmatrix} 1 \\ 2 \end{pmatrix}$ if the standard basis is more convenient to use. So, in a sense, the vector has existence independent of the "name" we use for it, where by "name" here I mean the column vector I choose to describe it.

Coordinate vectors are useful in another context. Recall that I showed you that the vectors

$$\vec{u} = \begin{pmatrix} -1 \\ 3 \\ 0 \end{pmatrix}, \quad \vec{v} = \begin{pmatrix} 2 \\ 0 \\ 1 \end{pmatrix} \tag{2.2.24}$$

together span a plane in $\mathbb{R}^3$. So every vector in the plane may be described as a linear combination of $\vec{u}$ and $\vec{v}$ by definition. Moreover, these vectors are linearly independent, so they have all the essential features of a basis! (See Fig. 2.6.) Define $\mathcal{B} = \{\vec{u}, \vec{v}\}$, and consider some vector $\vec{x}$ in the plane, say $\vec{x} = \begin{pmatrix} 1 \\ 3 \\ 1 \end{pmatrix}$. It's not hard to show that $\vec{x} = 1\vec{u} + 1\vec{v}$. With respect to the vectors in $\mathcal{B}$, we can describe $\vec{x}$ succinctly as

$$[\vec{x}]_{\mathcal{B}} = \begin{pmatrix} 1 \\ 1 \end{pmatrix}. \tag{2.2.25}$$

Notice that we need fewer components to describe $\vec{x}$ in terms of the basis $\mathcal{B}$ for the plane than we do in order to describe $\vec{x}$ in terms of the standard basis of $\mathbb{R}^3$! Coordinate vectors can then provide a more efficient way of describing vectors if we know they belong to special subsets of $\mathbb{R}^n$ possessing a basis[5].

[5]These special sets are called **subspaces** and I'll have more to say about them soon.

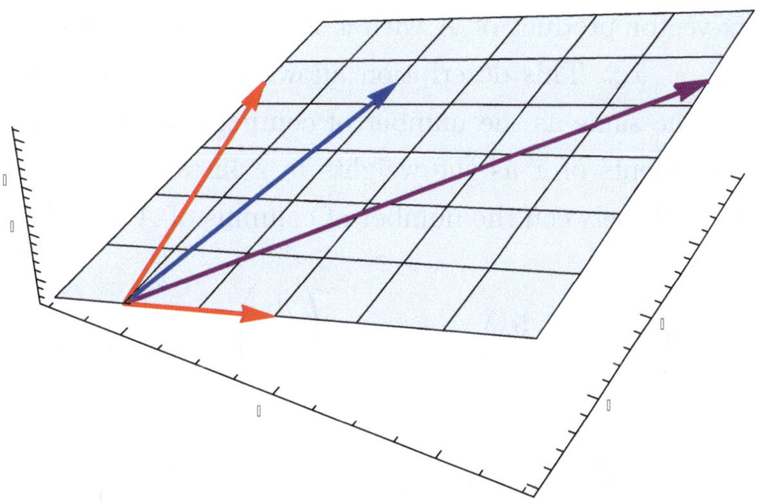

Figure 2.6: Plot of the plane $x_1 + \frac{1}{3}x_2 - 2x_3 = 0$ along with basis vectors $\mathcal{B} = \{\vec{u}, \vec{v}\}$ (red) and the linear combinations $\vec{x} = 1\vec{u} + 1\vec{v}$ (blue) and $\vec{y} = 1\vec{u} + 2\vec{v}$ (purple).

2.3 The Matrix-Vector Product

Another important concept in linear algebra stemming from the basic operations of vector addition and scalar multiplication is the so-called matrix-vector product. Matrix-vector products allow us to write linear systems of equations in a nice compact way; part of this section is devoted to understanding $A\vec{x} = \vec{b}$, which represents a linear system with unknowns listed in $\vec{x}$ and coefficient matrix A. We'll see that a consequence of writing the linear system of equations in the form $A\vec{x} = \vec{b}$ (together with the basic properties of the product) allow me to easily prove for you that any consistent linear system either has a single unique solution or an infinite family of solutions.

Definition: Let A be an $m \times n$ matrix and let $\vec{x} \in \mathbb{R}^n$.[6] Write A in terms of its columns as

$$A = (\vec{a}_1, \ldots, \vec{a}_n). \tag{2.3.1}$$

[6]It is important to note here that the number of columns of A has to be equal to the number of components of $\vec{x}$ for this definition to work!

Then,

$$A\vec{x} = x_1\vec{a}_1 + \cdots + x_n\vec{a}_n. \tag{2.3.2}$$

In words, the matrix-vector product of A with $\vec{x}$ is the linear combination of the columns of A with weights $x_1, \ldots, x_n$. This description allows us to see why we need the number of columns of A to be the same as the number of components of $\vec{x}$; otherwise, we couldn't naturally use the components of $\vec{x}$ as the weights in a linear combination of the $\vec{a}$'s since there would be a mismatch between the number of columns of A and the number of weights.

Example: Given $A = \begin{pmatrix} 1 & -1 & 5 \\ 2 & 3 & 0 \end{pmatrix}$ and $\vec{x} = \begin{pmatrix} 1 \\ 1 \\ 2 \end{pmatrix}$, let's compute $A\vec{x}$. The columns of A are

$$\vec{a}_1 = \begin{pmatrix} 1 \\ 2 \end{pmatrix}, \ \vec{a}_2 = \begin{pmatrix} -1 \\ 3 \end{pmatrix}, \ \vec{a}_3 = \begin{pmatrix} 5 \\ 0 \end{pmatrix}, \tag{2.3.3}$$

so

$$A\vec{x} = 1\begin{pmatrix} 1 \\ 2 \end{pmatrix} + 1\begin{pmatrix} -1 \\ 3 \end{pmatrix} + 2\begin{pmatrix} 5 \\ 0 \end{pmatrix} = \begin{pmatrix} 10 \\ 5 \end{pmatrix}. \tag{2.3.4}$$

In this example, we see that a 2×3 matrix and a vector in $\mathbb{R}^3$ multiply together to give us a vector in $\mathbb{R}^2$. This is an example of a general fact that follows directly from the definition: if A is $m \times n$ and $\vec{x} \in \mathbb{R}^n$, then $A\vec{x} \in \mathbb{R}^m$. We can now think of matrices as objects that take vectors and transform them into other vectors; i.e. we can think of A as defining a **function** of vectors. This perspective will profit us much in section 2.5.

It is not hard to show (based on the definition) that the matrix-vector product satisfies two important properties:

1. $A(\vec{x} + \vec{y}) = A\vec{x} + A\vec{y}$ (i.e. the multiplication is distributive).
2. $A(c\vec{x}) = c(A\vec{x})$ (i.e. scalars can be brought outside the matrix-vector product).

With this definition of matrix-vector product, we can write the most general linear system of equations as $A\vec{x} = \vec{b}$, where $\vec{x}$ is the vector made up of the unknown variables $x_1, \ldots, x_n$. I am now in a position to show that any consistent linear system necessarily has one or an infinite number of solutions.

The argument goes something like this: since the system is consistent, there is *at least one* solution in the solution set S. Say, for sake of argument, that S possesses two distinct solutions $\vec{x}, \vec{y} \in S$, $\vec{x} \neq \vec{y}$. Together, $\vec{x}$ and $\vec{y}$ define a line in $\mathbb{R}^n$, which is parametrically described as $t(\vec{x}-\vec{y})+\vec{y}$ for $t \in \mathbb{R}$. From the properties listed above, I can show that $t(\vec{x}-\vec{y})+\vec{y}$ is also a solution for every t!

$$\begin{aligned} A\big[t(\vec{x}-\vec{y})+\vec{y}\big] &= A\big[t(\vec{x}-\vec{y})\big]+A\vec{y} \\ &= tA(\vec{x}-\vec{y})+A\vec{y} \\ &= t(A\vec{x}-A\vec{y})+A\vec{y} \\ &= t\left(\vec{b}-\vec{b}\right)+\vec{b} \\ &= \vec{b}. \end{aligned} \tag{2.3.5}$$

Since t can be any real number, we have an infinite number of solutions in S if it contains more than one solution. Be sure you can see how each line in the argument follows from the previous line and the properties of the matrix-vector product.

Having the ability to write the most general linear system of equations as a **vector equation** $A\vec{x}=\vec{b}$ has enormous advantages, like making theorem proofs easier as I just demonstrated. But linear algebra is much more than the study of linear systems. It also includes the geometric properties of vectors, which may also be studied with matrix-vector products. In particular, the matrix-vector product allows us to define the **dot product**, which completely encodes the geometric content of vectors in $\mathbb{R}^n$[7].

The Dot Product

Earlier, I introduced the standard basis vectors $\hat{e}_j$. The hat in $\hat{e}_j$ is supposed to indicate unit length; i.e. the vector's arrow is one unit long. Intuitively, you have a good idea of what this means, but can we be more mathematically precise? The answer is yes: we can find formulas for not only the length of a given vector but also the angle between two nonzero vectors using the **dot product**. Informally, the dot product is a rule for multiplying two vectors $\vec{u}, \vec{v} \in \mathbb{R}^n$ to get a scalar, or number[8].

[7] Specifically, I am referring to Euclidean geometry. Other geometries are possible, but this topic is too large to undertake here.

[8] This is why the dot product is also sometimes referred to as the **scalar product**.

Before I can define the dot product, I need to introduce the operation of **matrix transpose**. The transpose operation effectively flips the matrix A across its **main diagonal**, the main diagonal consisting of the entries a_{11}, a_{22}, etc. Another way of looking at the transpose is that it forms the matrix A^T where the roles of rows and columns in A switch.

Definition: Let A be an $m \times n$ matrix. Then, the **matrix transpose** A^T is the $n \times m$ matrix defined by

$$\left(A^T\right)_{ij} = A_{ji}. \tag{2.3.6}$$

where, in general, B_{ij} is the (i,j)-component of the matrix B.

Example: Let $A = \begin{pmatrix} 1 & -1 & 5 \\ 2 & 3 & 0 \end{pmatrix}$. Then, $A^T = \begin{pmatrix} 1 & 2 \\ -1 & 3 \\ 5 & 0 \end{pmatrix}$.

Example: Let $\vec{x} = \begin{pmatrix} 1 \\ -1 \\ 4 \\ 2 \end{pmatrix}$. Then, $\vec{x}^{\,T} = \begin{pmatrix} 1 & -1 & 4 & 2 \end{pmatrix}$.[9]

The transpose operation enjoys many interesting and useful properties:

1. $\left(A^T\right)^T = A$.
2. $(A+B)^T = A^T + B^T$.
3. $(cA)^T = cA^T$.
4. $(AB)^T = B^T A^T$.

The last one is particularly interesting since we see that the transpose operation in a sense reverses the order of matrix multiplication: the transpose of the product is the product of the transposes **in reverse order**.[10] Most importantly for our purposes the transpose operation turns an $m \times n$ matrix into an $n \times m$ matrix. In particular, a column vector becomes a **row vector** (i.e. $1 \times n$ matrix) like in our last example.

[9] As a matter of practicality, I will use the transpose to write lengthy column vectors as row vectors to save paper!

[10] I haven't defined matrix products yet, so just keep this property in the back of your mind for now.

This is significant, because $\vec{u}^{\,T}\vec{v}$ is a well-defined matrix-vector product that is always 1×1 whenever $\vec{u}, \vec{v} \in \mathbb{R}^n$; it is a matrix with a single entry. In this case, we can forget about the ( ) and think of $\vec{u}^{\,T}\vec{v}$ as a single real number.

Definition: The **dot product** of $\vec{u}, \vec{v} \in \mathbb{R}^n$ is defined by

$$\vec{u} \cdot \vec{v} = \vec{u}^{\,T}\vec{v} = u_1 v_1 + \cdots + u_n v_n. \tag{2.3.7}$$

The dot product enjoys the following very nice properties that follow immediately from the definition and the properties of the matrix transpose: let $\vec{u}, \vec{v}, \vec{w} \in \mathbb{R}^n$ and let $c \in \mathbb{R}$. Then

1. $\vec{u} \cdot \vec{v} = \vec{v} \cdot \vec{u}$
2. $(\vec{u} + \vec{v}) \cdot \vec{w} = \vec{u} \cdot \vec{w} + \vec{v} \cdot \vec{w}$
3. $(c\,\vec{u}) \cdot \vec{v} = c\,\vec{u} \cdot \vec{v} = \vec{u} \cdot (c\,\vec{v})$
4. $\vec{u} \cdot \vec{u} \geq 0$ and $\vec{u} \cdot \vec{u} = 0$ if and only if $\vec{u} = 0$.

The last property gives us a clue as to how to define the length of a vector $\vec{u}$: the length must always be a positive number or zero.

Definition: The **length** or **norm** of a vector $\vec{u} \in \mathbb{R}^n$ is given by $||\vec{u}|| = \sqrt{\vec{u} \cdot \vec{u}}$.

In fact, if you consider $\vec{u} \in \mathbb{R}^2$, then $||\vec{u}|| = \sqrt{u_1^2 + u_2^2}$, which is exactly the distance of the point (u_1, u_2) from the origin $(0, 0)$ by the Pythagorean theorem. This in turn is the length of the arrow pointing from the origin to the point (u_1, u_2). So my definition is entirely reasonable.

Exercise: Show that the length of each standard basis vector $\hat{e}_j$ is, in fact, 1, as claimed.

The dot product gives us more than just the lengths of individual vectors; it also allows us to calculate geometrically interesting quantities like the angle between two vectors:

Definition: The **angle** between vectors $\vec{u}, \vec{v} \in \mathbb{R}^n$ with $\vec{u}\ ,\ \vec{v} \neq \vec{0}$ is the angle $\theta \in [0, \pi]$

satisfying

$$\cos\theta = \frac{\vec{u}\cdot\vec{v}}{||\vec{u}||||\vec{v}||}. \tag{2.3.8}$$

This equation can be derived in $\mathbb{R}^2$ using the Law of Cosines, but I will simply state the formula here without proof.

Example: Compute the angle between $\vec{u} = \begin{pmatrix} 12 \\ 3 \\ -5 \end{pmatrix}$, $\vec{v} = \begin{pmatrix} 2 \\ -3 \\ 3 \end{pmatrix}$. Using the definition of the dot product, we see that

$$\vec{u}\cdot\vec{v} = 12 \times 2 + 3 \times (-3) + (-5) \times 3 = 0. \tag{2.3.9}$$

Using the formula for the angle between two vectors, we see that $\cos(\theta) = 0$, so $\theta = \frac{\pi}{2}$. This is an example of a pair of **orthogonal vectors**:

Definition: Two vectors $\vec{u}$ and $\vec{v}$ in $\mathbb{R}^n$ are **orthogonal** to each other if $\vec{u}\cdot\vec{v} = 0$. In other words, the angle between the vectors is 90° (or $\frac{\pi}{2}$ radians).

Example: Compute the angle between $\vec{u} = \begin{pmatrix} 2 \\ 0 \end{pmatrix}$, $\vec{v} = \begin{pmatrix} 3 \\ 3 \end{pmatrix}$. Using the angle formula:

$$\cos(\theta) = \frac{\vec{u}\cdot\vec{v}}{||\vec{u}||||\vec{v}||} = \frac{2\times 3 + 0 \times 3}{\sqrt{2^2+0^2}\sqrt{3^2+3^2}} = \frac{1}{\sqrt{2}}, \tag{2.3.10}$$

so $\theta = \frac{\pi}{4}$ or 45°.

2.4 Abstract Vector Spaces and Subspaces

As I have defined them, column vectors enjoy the following set of rather obvious properties:

1. $\vec{u} + \vec{v} = \vec{v} + \vec{u}$
2. $\vec{u} + (\vec{v} + \vec{w}) = (\vec{u} + \vec{v}) + \vec{w}$
3. $\vec{u} + \vec{0} = \vec{u}$

4. $\vec{u} + (-\vec{u}) = \vec{0}$

5. $c(\vec{u} + \vec{v}) = c\vec{u} + c\vec{v}$

6. $(c + d)\vec{u} = c\vec{u} + d\vec{u}$

7. $c(d\vec{u}) = cd\vec{u}$

8. $1\vec{u} = \vec{u}$

These properties follow immediately from the definitions of scalar multiplication and vector addition (as you should verify on your own). So what? Well, it turns out that the powerful theorems and results of linear algebra rely entirely on these properties of vectors. They are, in a very important sense, the most fundamental properties of the building blocks of linear algebra.

This observation allows us to **abstract** these properties to an arbitrary set of objects. In more down-to-earth terms, this basically means we broaden our perspective to other collections of objects sharing the same properties as column vectors so that we can apply what we know about the linear algebra of column vectors to an even larger array of problems. An example or two should help clarify what I mean.

Example: Suppose I define the set P_n to be the set of all polynomials of degree n or less with real coefficients. In other words,

$$P_n = \{p(t) = p_n t^n + p_{n-1}t^{n-1} + \cdots + p_0 | p_j \in \mathbb{R}\}. \tag{2.4.1}$$

We know that is possible to add two elements of P_n together

$$\begin{aligned} p(t) + q(t) &= \left(p_n t^n + p_{n-1}t^{n-1} + \cdots + p_0\right) + \left(q_n t^n + q_{n-1}t^{n-1} + \cdots + q_0\right) \\ &= (p_n + q_n)\, t^n + \cdots + (p_0 + q_0) \end{aligned} \tag{2.4.2}$$

and multiply elements of P_n by scalars

$$cp(t) = (cp_n)\, t^n + \cdots + (cp_0) \tag{2.4.3}$$

in such a way that all eight properties above hold with obvious analogues. In this sense, P_n shares all of the essential features of vectors in $\mathbb{R}^n$.

Example: If I denote the set of continuous functions on the interval $[a, b]$ by $\mathcal{C}([a, b])$, then I can add two elements $f, g \in \mathcal{C}([a, b])$ to get a new element in $\mathcal{C}([a, b])$

$$(f + g)(t) = f(t) + g(t) \tag{2.4.4}$$

and multiply any element in $\mathcal{C}([a, b])$ by a constant

$$(cf)(t) = cf(t) \tag{2.4.5}$$

so that the set $\mathcal{C}([a, b])$, like P_n, shares all of the essential features of column vectors in $\mathbb{R}^n$.

Definition: An **abstract vector space**, or **vector space** for short, is a nonempty set V of objects, called **vectors**, on which are defined two operations, called **addition** and **scalar multiplication**, subject to the eight axioms listed below. The axioms must hold for all vectors $\boldsymbol{u}, \boldsymbol{v}, \boldsymbol{w}$ in V and for all scalars c and d and the set V must be **closed** under the two operations of scalar multiplication and addition:

For all $\boldsymbol{u}, \boldsymbol{v}, \boldsymbol{w} \in \mathbb{R}^n$ and all scalars c and d:

1. $\boldsymbol{u} + \boldsymbol{v} = \boldsymbol{v} + \boldsymbol{u}$
2. $\boldsymbol{u} + (\boldsymbol{v} + \boldsymbol{w}) = (\boldsymbol{u} + \boldsymbol{v}) + \boldsymbol{w}$
3. There is a $\mathbf{0} \in V$ such that $\boldsymbol{u} + \mathbf{0} = \boldsymbol{u}$
4. $\boldsymbol{u} + (-\boldsymbol{u}) = \mathbf{0}$
5. $c(\boldsymbol{u} + \boldsymbol{v}) = c\boldsymbol{u} + c\boldsymbol{v}$
6. $(c + d)\boldsymbol{u} = c\boldsymbol{u} + d\boldsymbol{u}$
7. $c(d\boldsymbol{u}) = cd\boldsymbol{u}$
8. $1\boldsymbol{u} = \boldsymbol{u}$

To distinguish abstract vectors from column vectors when the distinction is necessary, I use boldfaced symbols for abstract vectors and retain the overhead arrow notation for geometric vectors in $\mathbb{R}^n$. You should note that, under the terms of the definition, $\mathbb{R}^n$ with the usual operations of vector addition and scalar multiplication is an abstract vector space as well.

The most prominent application of abstract vector spaces is in the area of quantum theory, which governs the physics of the very, very small and is (as far as we know) a complete description of the how the universe works at a fundamental level. Modern technology like computers and lasers are based on the mathematics of abstract vector spaces and the axioms of quantum theory. Abstract vector spaces also enter into the area of Fourier analysis, a special topic I will present after we have a firm understanding of inner product spaces.

We've already encountered some basic vector spaces besides the ones described above and the more familiar $\mathbb{R}^n$, though I did not refer to them as vector spaces at the time of their introduction. In fact, planes in $\mathbb{R}^3$ like the one we saw before ($x_1 + \frac{1}{3}x_2 - 2x_3 = 0$ spanned by $\vec{u} = \begin{pmatrix} -1 & 3 & 0 \end{pmatrix}^T$ and $\vec{v} = \begin{pmatrix} 2 & 0 & 1 \end{pmatrix}^T$) are vector spaces in their own right. More generally,

Theorem: Given any collection of vectors $\vec{u}_1, \ldots, \vec{u}_p \in \mathbb{R}^n$, $\text{Span}\{\vec{u}_1, \ldots, \vec{u}_p\}$ is a vector space.

The proof just involves checking that all of the axioms of a vector space hold (which you should do on you own). As you can see from Fig. 2.6, such planes (and spanning sets in general) exist within a larger vector space $\mathbb{R}^n$. Since they are vector spaces and also live inside a larger vector space, they are referred to as **vector subspaces**.

Definition: A **vector subspace** of vector space V is any nonempty subset W of V satisfying:

1. The zero vector $\mathbf{0}$ is in W.
2. For each $\boldsymbol{w}_1$ and $\boldsymbol{w}_2$ in W, the sum $\boldsymbol{w}_1 + \boldsymbol{w}_2$ is in W.
3. For each $\boldsymbol{w}$ in W and each scalar c, the vector $c\boldsymbol{w}$ is in W.

In words, a nonempty subset of a vector space is a subspace if it is **closed** under addition and scalar multiplication; i.e. these operations don't carry vectors in W outside of W.

Example: Recall the polynomial space P_n. The collection of polynomials $p(t)$ satisfying $p(0) = 0$ is a subspace; it is not empty, because $p(t) = 0$ belongs to the set. Furthermore, if $p(0) = 0$ and $q(0) = 0$, then $cp(0) = 0$ and $p(0) + q(0) = 0$.

In the subspace represented by the plane in Fig. 2.6, we know that $\vec{u}$ and $\vec{v}$ constitute a **basis** for the space; all vectors in the space can be written in one and only one way as a linear combination of $\vec{u}$ and $\vec{v}$. The notion of a basis generalizes nicely:

Definition: A **basis** for a subspace W of a vector space V is a linearly independent set in W that spans W.

In this text, I will only focus on subspaces that have finite bases.[11] It turns out that if a subspace W has a finite basis, then any other basis shares exactly the same number of basis elements.

This is an important observation. A space like $\mathbb{R}^n$ can have an infinite number of bases. Just imagine taking a given set of coordinate axes and then rotating the whole rigid structure by various angles to get new coordinate systems. But, this observation tells us that no matter what, the *number* of basis *elements* must be fixed. This number has a special name:

Definition: The **dimension** of a nonzero subspace W, written $\dim W$, is the number of vectors in any basis for W. The dimension of the zero subspace $\{\mathbf{0}\}$ is defined to be 0.

Of course, $\mathbb{R}^n$ has dimension n as you would expect based on the intuitive notion of dimension. This follows from the definition of dimension since there are n standard basis vectors $\hat{e}_j$. Focusing on subspaces W with a finite dimension simplifies our work immensely. Moreover, it is possible to represent such subspaces in terms of a more familiar set of column

[11]It is certainly possible to have a vector subspace or space with an infinite number of basis vectors. Examples of such spaces arise in real analysis, partial differential equations, and theoretical physics. They share many properties in common with the vector spaces in this book, so there is no great loss for our purposes in focusing only on the finite case, which is much simpler to understand and work with.

vectors $\mathbb{R}^n$, which helps us to not only visualize such abstract vector subspaces in some cases but also enables us to encode them in computers for rapid calculation.

To see how this works, suppose first that W has $\mathcal{B} = \{\boldsymbol{w}_1, \ldots, \boldsymbol{w}_p\}$ as a basis so that W is p-dimensional. Then every vector $\boldsymbol{w} \in W$ can be uniquely written as $\boldsymbol{w} = c_1\boldsymbol{w}_1 + \cdots + c_p\boldsymbol{w}_p$. The weights c_j are real numbers entirely specifying $\boldsymbol{w}$. If this seems familiar, it should, since I am essentially talking about **coordinates** again:

Definition: Suppose the set $\mathcal{B} = \{\boldsymbol{w}_1, \ldots, \boldsymbol{w}_p\}$ is a basis for a subspace W. For each $\boldsymbol{w} \in W$, the **coordinates of $\boldsymbol{w}$ relative to the basis $\mathcal{B}$** are the weights $c_1, \ldots, c_p$ such that

$$\boldsymbol{w} = c_1\boldsymbol{w}_1 + \cdots + c_p\boldsymbol{w}_p,$$

and the vector in $\mathbb{R}^p$

$$[\boldsymbol{w}]_{\mathcal{B}} = \begin{pmatrix} c_1 & \ldots & c_p \end{pmatrix}^T \tag{2.4.6}$$

is called the **coordinate vector of $\boldsymbol{w}$ (relative to $\mathcal{B}$)**.

Example: Let $\mathcal{B} = \{t\ ,\ t^2\}$ be a basis for the subspace Z_2 of P_2 consisting of all polynomials of degree 2 or less that vanish at $t = 0$. The coordinate vector of $2t^2 - \frac{1}{2}t \in Z_2$ with respect to $\mathcal{B}$ is

$$\left[2t^2 - \frac{1}{2}t\right]_{\mathcal{B}} = \begin{pmatrix} -\frac{1}{2} \\ 2 \end{pmatrix}. \tag{2.4.7}$$

If we make the general identification

$$\boldsymbol{w} \in W \leftrightarrow [\boldsymbol{w}]_{\mathcal{B}} \in \mathbb{R}^p \tag{2.4.8}$$

for some choice of basis, then we see that W is for all intents and purposes the same as $\mathbb{R}^p$![12] This correspondence allows us to model abstract vector subspaces of finite dimension on any basic computer.

To conclude this section, I will present two important subspaces directly related to a given matrix A that will be important for our future work involving eigenvalues and eigenvectors.

[12]Technically, we say that W and $\mathbb{R}^p$ are **isomorphic** ("equal shape or form").

Definition: The **null space** of an $m \times n$ matrix A, written Nul A, is the set of all solutions of the homogeneous equation $A\vec{x} = \vec{0}$. In set notation, Nul $A = \{\vec{x} \in \mathbb{R}^n | A\vec{x} = \vec{0}\}$.

It is a simple matter to show that Nul A is, in fact, a vector subspace: first, clearly $\vec{0} \in$ Nul A. Since matrix-vector multiplication is distributive and scalars can be brought outside matrix-vector products, we also have closure of Nul A under vector addition and scalar multiplication.

For any matrix A, we may find a basis $\mathcal{B}$ for the null space by parametrically describing solutions to $A\vec{x} = \vec{0}$. I'll present an example to give a feel for how this works.

Example: Find an explicit description of Nul A, by listing vectors that span the null space for

$$A = \begin{pmatrix} 1 & 2 & 4 & 0 \\ 0 & 1 & 3 & -2 \end{pmatrix}. \tag{2.4.9}$$

Solutions to $A\vec{x} = \vec{0}$ can be obtained by row reducing

$$A = \left(\begin{array}{cccc|c} 1 & 2 & 4 & 0 & 0 \\ 0 & 1 & 3 & -2 & 0 \end{array}\right)$$

to the rref

$$A = \left(\begin{array}{cccc|c} 1 & 0 & -2 & 4 & 0 \\ 0 & 1 & 3 & -2 & 0 \end{array}\right).$$

The first equation tells us that $x_1 = 2x_3 - 4x_4$, and the second equation tells us that $x_2 = -3x_3 + 2x_4$ with absolutely no constraints on x_3 or x_4. These variables are free to take on any possible real value. If we assemble all of the components of the solution vector $\vec{x}$, we have

$$\vec{x} = \begin{pmatrix} x_1 \\ x_2 \\ x_3 \\ x_4 \end{pmatrix} = \begin{pmatrix} 2x_3 - 4x_4 \\ -3x_3 + 2x_4 \\ x_3 \\ x_4 \end{pmatrix} = x_3 \begin{pmatrix} 2 \\ -3 \\ 1 \\ 0 \end{pmatrix} + x_4 \begin{pmatrix} -4 \\ 2 \\ 0 \\ 1 \end{pmatrix}. \tag{2.4.10}$$

It follows immediately that we can write the solution set S as

$$S = \text{Span}\left\{\begin{pmatrix} 2 & -3 & 1 & 0 \end{pmatrix}^T, \begin{pmatrix} -4 & 2 & 0 & 1 \end{pmatrix}^T\right\} \tag{2.4.11}$$

since x_3 and x_4 are arbitrary, and we can take them to be free weights, or **free parameters** in the linear combination. As an exercise, demonstrate that the pair of spanning vectors are, in fact, linearly independent. It then follows that we have found a basis for the null space of A!

A few comments are in order. First, we saw that system of equations was not enough to completely fix all of the variables. Above, I wrote x_1 and x_2 in terms of x_3 and x_4 and claimed that x_3 and x_4 could take on any real value. You might wonder why I didn't solve x_3 and x_4 in terms of x_1 and x_2 and let x_1 and x_2 take on any real value. Mathematically, there would be nothing wrong in doing so. However, I followed a **convention**, or agreement among mathematicians, that makes the following choice:

Convention: Let U be the reduced row echelon form of A, the coefficient matrix of a linear system $A\vec{x} = \vec{b}$. Every column containing a leading one corresponds to a variable that is called a **basic variable**; columns not possessing a leading one correspond to **free variables**. When free variables (which take on all possible real values) are present, the convention is to express all basic variables entirely in terms of free variables.

I will adhere to this convention throughout the book. Second, the following procedure illustrated in the example above always works to find a basis for the null space of a given matrix:

1. Set up the augmented matrix $(A|\vec{0})$, and find the reduced row echelon form of $(A|\vec{0})$.
2. Write basic variables in terms of free variables.
3. Assemble the solution vector $\vec{x}$, and write it as a linear combination with weights given by the free variables.
4. The basis for Nul A is formed from the columns of this combination.

Another important subspace associated with a matrix is its **column space**. Unlike null space, the column space is defined explicitly via linear combinations of the columns of A. In other words, no calculations are required to specify a spanning set for the column space, but some work is required to find a *basis*.

Definition: The **column space** of an $m \times n$ matrix A, written Col A, is the set of all linear combinations of the columns of A. Specifically, if $A = (\vec{a}_1, \ldots, \vec{a}_n)$, then

$$\text{Col } A = \text{Span}\{\vec{a}_1, \ldots, \vec{a}_n\}$$

When finding a basis for Col A for some matrix A, start with the spanning set consisting of all columns of A. Then by definition, $\text{Col}A = \text{Span}\{\vec{a}_1, \ldots, \vec{a}_n\}$. In general, this spanning set need not be linearly independent. Suppose $A = (\vec{a}_1, \ldots, \vec{a}_n)$. In order to outline how to find a basis for Col A, I need to introduce the notion of **pivots**:

Definition: Let U be the reduced row echelon form of A. A column of A for which the corresponding column in U contains a leading one is called a **pivot column**, and the position in A for which the corresponding position in U contains a leading one is called a **pivot position** or just a **pivot**.

It turns out that the pivot columns of a matrix form a basis for the column space! The argument goes something like this: suppose $\vec{a}_1, \ldots, \vec{a}_n$ are the columns of matrix A. Any nontrivial solution to $A\vec{x} = \vec{0}$ gives us a set of nontrivial weights such that $x_1\vec{a}_1 + \cdots + x_n\vec{a}_n = \vec{0}$. If the only solution is $\vec{x} = \vec{0}$, then the columns are linearly independent and we're done. Suppose, on the other hand, that there are an infinite number of nontrivial solutions. After relabeling the columns (or shuffling them around if you like that idea better), we can assume that columns $\vec{a}_1, \ldots, \vec{a}_k$ are basic columns corresponding to basic variables $x_1, \ldots, x_k$, and columns $\vec{a}_{k+1}, \ldots, \vec{a}_n$ correspond to free variables $x_{k+1}, \ldots, x_n$.

Since x_{k+1}, etc. (the remaining variables) are free, I can choose any value I please for each of these variables. The basic variables are completely determined by these choices. If I pick $x_{k+1} = -1$, $x_{k+2} = 0$, $\ldots$ $x_n = 0$, then the basic variables $x_1, \ldots, x_k$ are completely determined by this choice, and we would have the relationship

$$x_1\vec{a}_1 + \cdots + x_k\vec{a}_k - \vec{a}_{k+1} = \vec{0} \tag{2.4.12}$$

or

$$x_1\vec{a}_1 + \cdots + x_k\vec{a}_k = \vec{a}_{k+1} \tag{2.4.13}$$

demonstrating that $\vec{a}_{k+1}$ is a linear combination of $\vec{a}_1, \ldots, \vec{a}_k$. The same argument shows that all of the other columns corresponding to free variables are linear combinations of the basic columns. In short, I have shown that

$$\text{Col } A = \text{Span } \{\vec{a}_1, \ldots, \vec{a}_k\}; \tag{2.4.14}$$

i.e. the basic columns span the column space.

If I can show that the basic columns are linearly independent, then I can conclude that the basic columns form a basis. If there is a set of weights $c_1, \ldots, c_k$ (not all 0) such that

$$c_1\vec{a}_1 + \cdots + c_k\vec{a}_k = \vec{0}, \tag{2.4.15}$$

then there is a nontrivial solution to $A\vec{x} = \vec{0}$ of the form

$$\vec{x} = \begin{pmatrix} c_1 & c_2 & \ldots & c_k & 0 & \ldots & 0 \end{pmatrix}^T, \tag{2.4.16}$$

meaning all the free variables are set to 0. Since the basic variables are written in terms of the free variables, it follows that all of the basic variables are 0 as well, contradicting the assumption I made above. So the basic columns are linearly independent, and they span the column space. By definition, they form a basis for Col A.

Therefore, if you want a basis for Col A, you need only determine the pivot positions (through row reduction or some other argument), then select those vectors from the list $\vec{a}_1, \ldots, \vec{a}_n$ that contain a pivot position.

Example: Compute a basis for the matrix $A = \begin{pmatrix} 4 & 5 & 9 & -2 \\ 6 & 5 & 1 & 12 \\ 3 & 4 & 8 & -3 \end{pmatrix}$.

Applying the Gauss-Jordan algorithm, the rref of the matrix is $U = \begin{pmatrix} 1 & 0 & -4 & 7 \\ 0 & 1 & 5 & -6 \\ 0 & 0 & 0 & 0 \end{pmatrix}$.

According to the definitions above, the first and second columns are pivot columns, so $\vec{a}_1 = \begin{pmatrix} 4 \\ 6 \\ 3 \end{pmatrix}$ and $\vec{a}_2 = \begin{pmatrix} 5 \\ 5 \\ 4 \end{pmatrix}$ form a basis for Col A.

To conclude this discussion, I will now present a theorem connecting the dimensions of the column space and the null space. The theorem is fundamental to proving theorems and has many applications in the theory of linear algebra. I include it here mostly for completeness and to establish a connection between the column and null spaces of a matrix.

Defintion: The **rank** of a matrix A, denoted rank A, is the dimension of the column space of A. The **nullity** of a matrix A, denoted null A, is the dimension of the null space.

Theorem (Rank-Nullity Theorem): If a matrix A has n columns, then

$$\text{rank } A + \text{null } A = n. \tag{2.4.17}$$

One important consequence of this theorem is that the dimension of the column space of A (i.e., the rank of A or, more specifically, the **column rank**) is the same as the dimension of the **row space of A**. The row space of A is the vector space spanned by the rows of A viewed as vectors. The dimension of the row space is often referred to as the **row rank**.

2.5 Linear Transformations on Vector Spaces

In the next chapter, I will explore matrices in greater detail. There, I will show you how matrices gain a life of their own independent of any sort of linear system of equations. In other words, matrices are interesting in and of themselves. The way to this perspective begins by thinking of matrices as functions of vectors.

I've already shown you how matrices and vectors multiply together. If A is $m \times n$ and $\vec{x} \in \mathbb{R}^n$, then $A\vec{x}$ is a vector in $\mathbb{R}^m$. In effect, the matrix-vector product takes an n-component vector and turns it into an m-component vector. This is precisely what a function does!

Definition: A function f is a relation between two sets of objects D (called the domain) and C (called the codomain) such that every object in D is assigned exactly one object in C.

As a shorthand, mathematicians usually write $f : D \to C$ to indicate that f is a function from D to C. If $d \in D$, then we say that $c = f(d)$ is the **image** of d under f. The purpose of stating the codomain is to indicate what type of output the function f gives. It is not always the case that the **range** of f (i.e. the set of all images of f) is the same as the codomain:

Example: Let $f : \mathbb{R} \to \mathbb{R}$ be defined by $f(x) = x^2$. The codomain is $\mathbb{R}$, since the output of f is a real number, but the range of f is just $[0, \infty)$ or just the nonnegative real numbers.

In the above definition, the domain and codomain are general nonempty sets. In this book, I will limit our attention just to the situation where the domain and codomain are vector spaces. For the time being, assume $D = \mathbb{R}^n$ and $C = \mathbb{R}^m$. If A is $m \times n$, then I can define a function $T : \mathbb{R}^n \to \mathbb{R}^m$ as follows:

$$T(\vec{x}) = A\vec{x}. \tag{2.5.1}$$

Since the operation of matrix-vector multiplication satisfies $A(\vec{x} + \vec{y}) = A\vec{x} + A\vec{y}$ and $A(c\vec{x}) = cA\vec{x}$, we see that $T(\vec{x} + \vec{y}) = T(\vec{x}) + T(\vec{y})$ and $T(c\vec{x}) = cT(\vec{x})$. In fact, these are the defining properties of a function, or transformation, of vectors that preserves the linear structure of the vector space.

Definition: A **linear transformation** $T : V \to W$ between vector spaces V and W is a function satisfying

1. $T(\boldsymbol{x} + \boldsymbol{y}) = T(\boldsymbol{x}) + T(\boldsymbol{y})$

2. $T(c\boldsymbol{x}) = cT(\boldsymbol{x})$

for all $\boldsymbol{x}, \boldsymbol{y} \in V$ and $c \in \mathbb{R}$.

For abstract vector spaces, this is one natural way of generalizing (or abstracting) the idea of matrix. The reason is simple: not only do matrices define linear transformations, but linear transformations $T : \mathbb{R}^n \to \mathbb{R}^m$ can always be represented by matrices. So matrices are, in a sense, entirely equivalent to linear transformations $T : \mathbb{R}^n \to \mathbb{R}^m$ between Euclidean spaces. Let's see how this works.

As I demonstrated earlier, any vector $\vec{x} \in \mathbb{R}^n$ has a unique representation in terms of the standard basis vectors $\hat{e}_j$: $\vec{x} = x_1\hat{e}_1 + \cdots + x_n\hat{e}_n$. Let $T : \mathbb{R}^n \to \mathbb{R}^m$ be some linear transformation from $\mathbb{R}^n$ to $\mathbb{R}^m$. Then

$$\begin{aligned} T(\vec{x}) &= T(x_1\hat{e}_1 + \cdots + x_n\hat{e}_n) \\ &= T(x_1\hat{e}_1) + \cdots + T(x_n\hat{e}_n) \\ &= x_1T(\hat{e}_1) + \cdots + x_nT(\hat{e}_n), \end{aligned} \tag{2.5.2}$$

all following from the linearity of T. The vectors $T(\hat{e}_j) \in \mathbb{R}^m$ are just the images of the standard basis vectors and are m-component column vectors. In the last line, we see that we have a linear combination of the vectors $T(\hat{e}_j)$ with weights x_j. But, this is precisely the matrix-vector product of A and $\vec{x}$ if I define

$$A = (T(\hat{e}_1), \ldots, T(\hat{e}_n)) \; ! \tag{2.5.3}$$

I have established the following theorem:

Theorem: Let $T : \mathbb{R}^n \to \mathbb{R}^m$ be a linear transformation and define the $m \times n$ matrix $A = (T(\hat{e}_1), \ldots, T(\hat{e}_n))$ called the **standard matrix representation of $\boldsymbol{T}$**. Then $T(\vec{x}) = A\vec{x}$ for all $\vec{x} \in \mathbb{R}^n$.

In words, if you are given a linear transformation between Euclidean spaces, then you can always construct a matrix representing that transformation; simply find all of the images of the standard basis vectors under T, and place them in order in a matrix. Viola!

Example: In some situations, you may only be given a verbal description of how T changes a vector $\vec{x} \in \mathbb{R}^n$. For instance, suppose a vector $\vec{x} \in \mathbb{R}^2$ is rotated by a fixed positive angle θ counterclockwise. It is not hard to argue that this transformation is linear, so I'll leave it to you as an exercise. According to the above theorem, we can construct a matrix representation A for T if we know how the standard basis vectors $\hat{e}_1$ and $\hat{e}_2$ transform. I work this out using basic trigonometry in Figure 2.7.

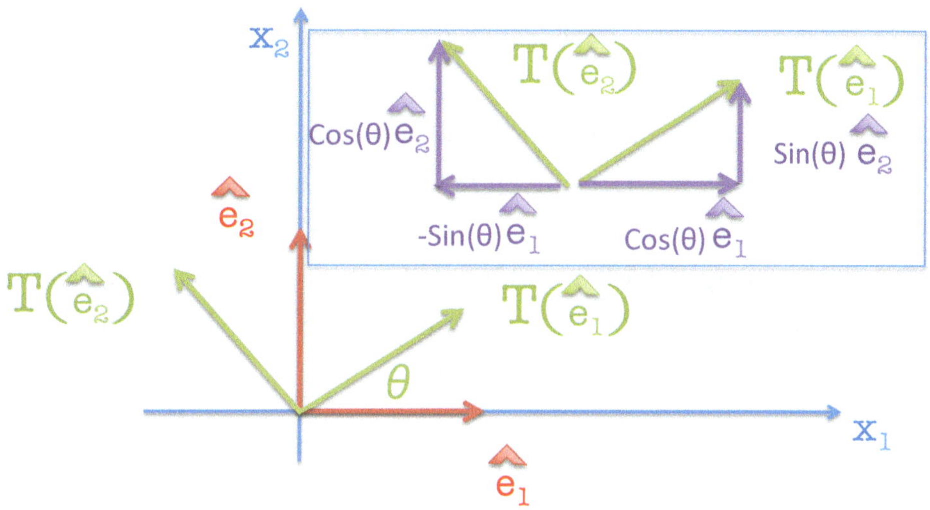

Figure 2.7: A rotation counterclockwise around the origin in $\mathbb{R}^2$ can be desribed as a linear transformation T (in green). The matrix representation of T can be obtained by finding the images (in green) of $\hat{e}_1$ and $\hat{e}_2$ (in red). The inset shows the basic trigonometry needed to find the components of $T(\hat{e}_1)$ and $T(\hat{e}_2)$ (in purple).

In short,

$$T(\hat{e}_1) = \begin{pmatrix} \cos(\theta) \\ \sin(\theta) \end{pmatrix}, \; T(\hat{e}_2) = \begin{pmatrix} -\sin(\theta) \\ \cos(\theta) \end{pmatrix}, \tag{2.5.4}$$

so

$$A = (T(\hat{e}_1) \, , \, T(\hat{e}_2)) = \begin{pmatrix} \cos(\theta) & -\sin(\theta) \\ \sin(\theta) & \cos(\theta) \end{pmatrix}. \tag{2.5.5}$$

In other situations, you may be given an explicit formula for T:

Example: Let $T : \mathbb{R}^4 \to \mathbb{R}^2$ be a linear transformation defined by $T(x_1, x_2, x_3, x_4) = (x_1 + x_3, x_2 - x_4)$. It is not too hard to see why this transformation is linear: you can show this directly by using the definition of a general linear transformation. The matrix representation is determined by the images of all the standard basis vectors of $\mathbb{R}^4$ under T:

$$T(\hat{e}_1) = \begin{pmatrix} 1 \\ 0 \end{pmatrix}, \; T(\hat{e}_2) = \begin{pmatrix} 0 \\ 1 \end{pmatrix}, \; T(\hat{e}_3) = \begin{pmatrix} 1 \\ 0 \end{pmatrix}, \; T(\hat{e}_4) = \begin{pmatrix} 0 \\ -1 \end{pmatrix}, \tag{2.5.6}$$

so

$$A = \begin{pmatrix} 1 & 0 & 1 & 0 \\ 0 & 1 & 0 & -1 \end{pmatrix}. \tag{2.5.7}$$

Now that you have a fairly good understanding of linear transformations between Euclidean spaces and how they relate to matrices, I can reformulate the idea of the column space of A in the language of functions. Specifically, it is true that Col A is the range of $T(\vec{x}) = A\vec{x}$. Indeed, $\vec{b} \in \mathbb{R}^m$ is in the **range** of $T(\vec{x}) = A\vec{x}$ if and only if $A\vec{x} = \vec{b}$ has at least one solution, which is equivalent to writing $\vec{b} = x_1\vec{a}_1 + \ldots x_n\vec{a}_n$ for some choice of weights $x_1, \ldots, x_n$. This, of course, means that $\vec{b}$ is in the span of the columns of A.

Just as linear transformations $T : V \to W$ allow us to generalize the notion of the matrix to abstract vector spaces, the range of T generalizes the notion of the column space; i.e. it is the subspace made up of all images of T. While I'm at it, I can write down a generalization for the null space as well:

Definition: The **kernel** of the linear transformation $T : V \to W$ is the subspace of all vectors in V annihilated by T:

$$\text{Ker } T = \{\boldsymbol{v} \in V | \; T(\boldsymbol{v}) = \boldsymbol{0}\}. \tag{2.5.8}$$

Example: The space of polynomials of degree no larger than n, P_n, is a vector space. We can define a linear transformation $D : P_n \to P_{n-1}$ for $n \geq 1$ by taking the derivative $D(p(t)) = \frac{dp}{dt} = p'(t)$. We know this transformation is linear from the basic rules of calculus. The kernel consists of all polynomials sent to 0 under the derivative. These are just the constant polynomials, so

$$\text{Ker } D = \{p(t) = c \in P_n | c \in \mathbb{R}\}. \tag{2.5.9}$$

The kernel and the range determine if $T : V \to W$ is **one-to-one** and/or **onto**. In general, a function is **one-to-one** if $f(x_1) = f(x_2)$ means that $x_1 = x_2$. In other words, any element of the range is the image of a unique element of the domain. A function like $f(x) = x^2$ is not one-to-one because ± 1 are sent to 1 after we square, but a function like $f(x) = 2x - 3$ is one-to-one.

A function is **onto** if the range is the same as the codomain; i.e. every element of the codomain is the image of at least one element of the domain. So $f(x) = x^2$ is not onto because the negative numbers are outside the range, but $f(x) = 2x - 3$ is onto, because we can always solve $y = 2x - 3$ for x given any $y \in \mathbb{R}$.

From the definition, we see that $T : V \to W$ is onto if and only if the range is the same as W. It turns out that $T : V \to W$ is one-to-one (assuming T is linear) if and only if Ker $T = \{\mathbf{0}\}$. It's certainly the case that $\mathbf{0} \in$ Ker T, since T is linear: it has no choice but to send $\mathbf{0} \in V$ to $\mathbf{0} \in W$ (think about why!). Suppose Ker $T = \{\mathbf{0}\}$; then $\mathbf{0} \in V$ is the *only* vector to be sent to $\mathbf{0} \in W$. If it happens to be the case that $T(\boldsymbol{v}_1) = T(\boldsymbol{v}_2)$, then $T(\boldsymbol{v}_1 - \boldsymbol{v}_2) = \mathbf{0}$ by linearity, so $\boldsymbol{v}_1 - \boldsymbol{v}_2 = \mathbf{0}$ or $\boldsymbol{v}_1 = \boldsymbol{v}_2$. This establishes that T is one-to-one. If T is one-to-one, then since $\mathbf{0} \in V$ is taken to $\mathbf{0} \in W$ under T, it follows that Ker $T = \{\mathbf{0}\}$.

To end this discussion, let me emphasize that linear transformations can just be one-to-one, just onto, both, or neither.

Example: The linear transformation $T : \mathbb{R}^3 \to \mathbb{R}^2$ given by $T(x_1, x_2, x_3) = (x_1, x_2)$ is onto, but it is not one-to-one. In fact, given a point $(x_1, x_2) \in \mathbb{R}^2$, we know that $T(x_1, x_2, 0) = (x_1, x_2)$ (so it's onto) but also $T(x_1, x_2, c) = (x_1, x_2)$ for any choice of $c \in \mathbb{R}$ (so it's not one-to-one).

Example: The linear transformation $T : \mathbb{R}^2 \to \mathbb{R}^3$ given by $T(x_1, x_2) = (x_1, x_2, 0)$ is not onto (because any point of the form $(x_1, x_2, 1)$ is not in the range for instance), but it is one-to-one as you can easily show using the definition.

Example: The linear transformation $T : \mathbb{R}^n \to \mathbb{R}^m$ given by $T(\vec{x}) = \vec{0}$ is neither one-to-one nor onto.

Example: The linear transformation $T : \mathbb{R}^n \to \mathbb{R}^n$ given by $T(\vec{x}) = \vec{x}$ is both one-to-one and onto.

Example: Determine if the linear transformation defined by $A = \begin{pmatrix} -1 & 7 & 2 & 4 \\ 0 & 3 & 5 & 8 \\ 6 & 0 & -10 & 6 \end{pmatrix}$ is one-to-one. Determine if it is onto.

Given what I wrote above, $T(\vec{x}) = A\vec{x}$ will be one-to-one if and only if the kernel consists of just $\vec{0}$. The kernel of T is nothing but the null space of A. The null space in turn is the set of all vectors $\vec{x}$ satisfying $A\vec{x} = \vec{0}$. So T will be one-to-one if and only if the only solution to the system $A\vec{x} = \vec{0}$ is $\vec{0}$. You can easily demonstrate that this matrix has a free parameter by row reduction:

$$A \sim \begin{pmatrix} 1 & 0 & 0 & \frac{307}{102} \\ 0 & 1 & 0 & \frac{67}{102} \\ 0 & 0 & 1 & \frac{41}{34} \end{pmatrix} \tag{2.5.10}$$

Therefore, T is not one-to-one.

On the other hand, T will be onto if and only if the range of T is the same as $\mathbb{R}^3$, the codomain. In other words, T will be onto if and only if $A\vec{x} = \vec{b}$ always has a solution, regardless of what we pick for $\vec{b} \in \mathbb{R}^3$. From above, we know that A has a leading 1 in every row in its reduced row echelon form, so $A\vec{x} = \vec{b}$ always has a solution. Therefore, T is onto.

In the next chapter, I will present the basics of matrix operations. The best way to understand matrices on their own terms is by thinking of them as functions. We know how to add functions, subtract functions and multiply functions by constants. The same principles apply to matrices as I'll show you. But these operations are not the whole story. Under certain circumstances, it is possible to **compose** two functions together to obtain a new one. This process, in the context of matrices, leads us to a natural definition of matrix multiplication. Furthermore, some functions $f : D \to C$ (namely those that are one-to-one and onto) possess an **inverse function** $f^{-1} : C \to D$ that "undoes" the operation of f. Through this idea, I will introduce you to the notion of the **matrix inverse**.

2.6 Chapter 2 Exercises

1. For the following vector pairs $\vec{u}$ and $\vec{v}$, compute $-2\vec{u}$, $\vec{u}+2\vec{v}$ and $\frac{1}{2}\vec{v}$. Draw the results in the plane $\mathbb{R}^2$. For vector additions, use the Parallelogram Rule to aid your drawing.

   (a)
$$\vec{u} = \begin{pmatrix} 1 \\ -2 \end{pmatrix} \ , \ \vec{v} = \begin{pmatrix} 3 \\ -1 \end{pmatrix}$$

   (b)
$$\vec{u} = \begin{pmatrix} 5 \\ 0 \end{pmatrix} \ , \ \vec{v} = \begin{pmatrix} -2 \\ 3 \end{pmatrix}$$

2. Given two points P and Q in the Cartesian plane **not** on a vertical line, show that the **parametric description** $t\left(\vec{P}-\vec{Q}\right)+\vec{Q}\ ,\ t \in \mathbb{R}$ gives the familiar $y = mx + b$ form of the line once the parameter t is eliminated.

3. Find the parametric description of the line through P and Q if

   (a) $P = (-1, 0, 1)\ ,\ Q = (4, 1, 5)$

   (b) $P = (3, 0, 2, -1)\ ,\ (1, 5, 0, -7)$

   (c) $P = (0, 0, 0, 0, 0)\ ,\ Q = (1, -1, 1, -1, 1)$.

4. The solution set to the single linear equation $2x_1 - x_2 + 5x_3 = 0$ is a plane in $\mathbb{R}^3$ through the origin. Describe this plane as the span of a pair of vectors. Demonstrate explicitly that the vector pair is, in fact, linearly independent.

5. Determine if the following sets of vectors are linearly independent by performing row reduction on an appropriate augmented matrix:

   (a)
$$\vec{v}_1 = \begin{pmatrix} 1 \\ 1 \\ 2 \\ 4 \end{pmatrix} \ , \ \vec{v}_2 = \begin{pmatrix} -1 \\ 1 \\ 4 \\ 0 \end{pmatrix} \ , \ \vec{v}_3 = \begin{pmatrix} 4 \\ 0 \\ -4 \\ 8 \end{pmatrix} \ , \ \vec{v}_4 = \begin{pmatrix} 2 \\ 1 \\ 1 \\ 6 \end{pmatrix}$$

(b)

$$\vec{v}_1 = \begin{pmatrix} 1 \\ 0 \\ -1 \end{pmatrix}, \vec{v}_2 = \begin{pmatrix} 0 \\ -3 \\ 2 \end{pmatrix}, \vec{v}_3 = \begin{pmatrix} -1 \\ -2 \\ -1 \end{pmatrix}$$

6. For the sets of vectors in the previous problem, describe the corresponding span geometrically. The collection of vectors in (*b*) is, in fact, linearly independent. Express all standard basis vectors of $\mathbb{R}^3$ as linear combinations of these vectors.

7. Just by *inspection* alone, can you tell if the set of vectors $\vec{v}_1 = \begin{pmatrix} 1 \\ 0 \end{pmatrix}$, $\vec{v}_2 = \begin{pmatrix} 1 \\ -3 \end{pmatrix}$, $\vec{v}_3 = \begin{pmatrix} 5 \\ -2 \end{pmatrix}$ is linearly independent or not? If so, explain how you arrived at your answer without any calculation. How about the set of vectors $\vec{v}_1 = \begin{pmatrix} 4 \\ 0 \\ -1 \\ 5 \\ 7 \end{pmatrix}$, $\vec{v}_2 = \begin{pmatrix} 1 \\ 1 \\ 7 \\ 8 \\ -1 \end{pmatrix}$? Again, explain your reasoning.

8. I claimed that bases for $\mathbb{R}^n$ (or any vector space V for that matter) must share the same number of vectors. I defined this number to be the **dimension** of the space. In this exercise, I will walk you through an argument demonstrating my claim. We already know that the standard basis vectors $\hat{e}_1, \ldots, \hat{e}_n$ make up a basis[13] for $\mathbb{R}^n$ since they are linearly independent and span $\mathbb{R}^n$. It is enough to show that another basis $\mathcal{B}$ for $\mathbb{R}^n$ has exactly n elements. Explicitly, $\mathcal{B} = \{\vec{w}_1, \ldots, \vec{w}_m\}$, and we need to show that $m = n$. You know that m cannot be larger than n, because then the number of components of each vector would be less than the total number of vectors in $\mathcal{B}$, which means that the vectors in $\mathcal{B}$ are necessarily linearly dependent. This cannot be the case since a basis is necessarily a set of linearly independent vectors. Therefore, $m \leq n$. We just need to argue that $m < n$ cannot be true, and then the proof is complete.

   Suppose that $m < n$ is true. Demonstrate that it is possible to find a vector $\vec{x}$ that is **orthogonal** to each basis vector in $\mathcal{B}$ as follows: any vector in $\mathbb{R}^n$ can be written as

[13] Hence the name!

$\vec{x} = \sum_{i=1}^{n} x_i \hat{e}_i$, so $\vec{x} \cdot \vec{w}_j = \sum_{i=1}^{n} x_i (\hat{e}_i \cdot \vec{w}_j)$ by the properties of the dot product. The weights $x_1, \ldots, x_n$ are undetermined. If $\vec{x} \cdot \vec{w}_j = 0$ is true for all $j = 1, \ldots, m$, then the weights must be a solution to a linear system of equations with coefficients $\hat{e}_i \cdot \vec{w}_j$. This linear system is represented by a $m \times n$ coefficient matrix A and has $\vec{b} = \vec{0}$. Argue that, since $m < n$, there must be at least one *nonzero* solution. Therefore, there is at least one vector $\vec{x} \neq \vec{0} \in \mathbb{R}^n$ orthogonal to every basis vector in $\mathcal{B}$.

Since $\mathcal{B}$ is a basis for $\mathbb{R}^n$, it must be possible to write $\vec{x}$ as a linear combination of the vectors in $\mathcal{B}$. Specifically, $\vec{x} = c_1\vec{w}_1 + \cdots + c_m\vec{w}_m$. Since $\vec{x}$ is orthogonal to each of the $\vec{w}_j$, argue that $c_j = 0$ for all j. This implies that $\vec{x} = \vec{0}$, which is a contradiction.

This logical contradiction shows you that the original assumption must have been false. The original assumption, of course, was that $m < n$. Therefore, $m = n$ and any basis for $\mathbb{R}^n$ must have exactly n vectors. As you progress through the course, return to this problem, and see if any of the techniques I develop later for you help make the above claim easier to prove.

9. Define a linear transformation $T : \mathbb{R}^3 \to \mathbb{R}^2$ by $T(x_1, x_2, x_3) = (x_1 + 2x_2 - x_3, x_1 + 3x_3)$. Find the standard matrix representation of T, and determine if T is a one-to-one transformation. Determine if T is an onto transformation.

10. Find a basis for Col(A) and a basis for Nul(A) where $A = \begin{pmatrix} -2 & 4 & -2 & -4 \\ 2 & -6 & -3 & 1 \\ -3 & 8 & 2 & -3 \end{pmatrix}$. What is the rank of A? What is the nullity of A? Are your answers consistent with the Rank-Nullity Theorem? In the column space basis $\mathcal{B}$ you found above, what is $[\boldsymbol{x}]_{\mathcal{B}}$ for $\boldsymbol{x} = \begin{pmatrix} 2 \\ -4 \\ 5 \end{pmatrix}$?

11. Define the matrix $A = \begin{pmatrix} 3 & 0 & 1 \\ -2 & 0 & 0 \\ -2 & 0 & 0 \end{pmatrix}$. Find a basis for Nul$(A)$. What is the dimension of this space?

12. Show that the set $H = \text{Span}\{\cos(t), \sin(t)\}$ is a vector space contained within the larger

space $C([0, 2\pi])$, which in turn, is the space of all continuous functions on $[0, 2\pi]$. This set is **isomorphic** to $\mathbb{R}^2$ through the linear transformation $T : H \to \mathbb{R}^2$, $T(a\cos(t) + b\sin(t)) = \begin{pmatrix} a \\ b \end{pmatrix}$. From Calculus, it is not hard to see that the basic properties of the derivative $\frac{d}{dt}$ allow us to think of the derivative as a linear transformation on H. With this observation, construct a 2×2 matrix representing the derivative of elements of H with respect to t. Determine if the derivative is one-to-one on H by computing the **kernel**. Use the Rank-Nullity Theorem to show that the derivative is also onto.

13. The hyperbolic trigonometric functions $\cosh(t)$ and $\sinh(t)$ span a two-dimensional vector space contained within the larger space of continuous, real-valued functions. Furthermore, they are linearly independent. Define $\mathcal{B} = \{\cosh(t), \sinh(t)\}$, and compute $[\boldsymbol{x}]_{\mathcal{B}}$ for $\boldsymbol{x} = 2e^t - 3e^{-t}$. You may need the following formulae:

$$\cosh(t) = \frac{e^t + e^{-t}}{2}, \sinh(t) = \frac{e^t - e^{-t}}{2}. \tag{2.6.1}$$

14. Using the dot product, compute the angle between the following vector pairs:

(a) $\begin{pmatrix} 3 \\ 0 \\ 1 \end{pmatrix}, \begin{pmatrix} 0 \\ -1 \\ 1 \end{pmatrix}$

(b) $\begin{pmatrix} 5 \\ -1 \\ 2 \\ 0 \end{pmatrix}, \begin{pmatrix} 2 \\ 0 \\ 1 \\ -1 \end{pmatrix}$

15. This exercise is meant to be challenging and completes the theorem given in the text claiming that there is only one rref of a matrix. Given that a nonzero matrix M is row equivalent to a matrix R in rref, show that there are no other rref matrices row equivalent to M. In other words, if R and R' are rref matrices row equivalent to M, show that $R = R'$.

I will help guide you through a proof relying on the principle of induction. If M is an $m \times 1$ matrix, then the claim is easy to prove: M is a column vector with m components, at least one of which is nonzero. The Gauss-Jordan algorithm in this case involves using a nonzero entry to eliminate all of the other entries via row operations, scaling this nonzero entry to make a 1, and then finally moving this entry to the top-most position. Argue that there is only one possible column vector resulting from this process. The theorem is therefore true for the case of $n = 1$.

Suppose now that $n > 1$. Assume that M is row equivalent to two distinct $m \times n$ matrices R and R'. I will show you how to argue that this leads to a logical contradiction and so cannot be true. Write $M = (\tilde{M}|\vec{m}_n)$. Specifically, I'm isolating the last column of M with a partition and lumping the previous $n-1$ columns into a new $m \times (n-1)$ matrix called $\tilde{M}$. Using the principle of induction, I will assume that $\tilde{M}$ is row equivalent to a *unique* $m \times (n-1)$ matrix called $\tilde{R}$. Since the same row operations used to reduce M are used to reduce $\tilde{M}$, argue that $R = (\tilde{R}|\vec{u})$ and $R' = (\tilde{R}|\vec{u}\ ')$ for some vectors $\vec{u}$ and $\vec{u}\ '$. In other words, by induction, R and R' can only differ in their last columns.

To proceed, argue that any vector in $\text{Nul}(\tilde{R}|\vec{u})$ or $\text{Nul}(\tilde{R}|\vec{u}\ ')$ must have $x_n = 0$ if $\vec{u} \neq \vec{u}\ '$. Conclude that the last column of M is necessarily a pivot column and then argue that $\vec{u}$ must be the same as $\vec{u}\ '$, which is a contradiction. This concludes the inductive proof.

2.7 Chapter 2 References

If the reader wishes to explore the theory behind the theorems presented in this chapter or wishes to see more examples, I suggest consulting [5, 6]. The outline of the proof of the uniqueness of the rref of a nonzero matrix was largely inspired by [7]. A different proof is given in the appendix to [5].

Chapter 3

Matrices

In this section, matrices take center stage. I will show you how the basic operations of vector addition and scalar multiplication work for $m \times n$ matrices. I will also show you how thinking of matrices as linear transformations allows us to define (in some special cases) matrix products and matrix inverses, ideas which will prove very useful in later applications.

3.1 Matrix Addition and Scalar Multiplication

To begin, recall the definitions of vector addition and scalar multiplication in $\mathbb{R}^n$. We know vectors in $\mathbb{R}^n$ can be thought of as $n \times 1$ matrices; i.e. matrices with a single column. I defined vector addition and scalar multiplication component-wise:

$$\vec{u} + \vec{v} = \begin{pmatrix} u_1 \\ u_2 \\ \vdots \\ u_n \end{pmatrix} + \begin{pmatrix} v_1 \\ v_2 \\ \vdots \\ v_n \end{pmatrix} = \begin{pmatrix} u_1 + v_1 \\ u_2 + v_2 \\ \vdots \\ u_n + v_n \end{pmatrix}, \quad c\vec{u} = \begin{pmatrix} cu_1 \\ cu_2 \\ \vdots \\ cu_n \end{pmatrix}. \tag{3.1.1}$$

Why should we limit these definitions to just vectors? It is entirely natural to extend addition and scalar multiplication to matrices by defining them component-wise as well:

$$A + B = \begin{pmatrix} a_{11} & \dots & a_{1n} \\ \vdots & \vdots & \vdots \\ a_{m1} & \dots & a_{mn} \end{pmatrix} + \begin{pmatrix} b_{11} & \dots & b_{1n} \\ \vdots & \vdots & \vdots \\ b_{m1} & \dots & b_{mn} \end{pmatrix} = \begin{pmatrix} a_{11} + b_{11} & \dots & a_{1n} + b_{1n} \\ \vdots & \vdots & \vdots \\ a_{m1} + b_{m1} & \dots & a_{mn} + b_{mn} \end{pmatrix} \tag{3.1.2}$$

$$cA = c\begin{pmatrix} a_{11} & \dots & a_{1n} \\ \vdots & \vdots & \vdots \\ a_{m1} & \dots & a_{mn} \end{pmatrix} = \begin{pmatrix} ca_{11} & \dots & ca_{1n} \\ \vdots & \vdots & \vdots \\ ca_{m1} & \dots & ca_{mn} \end{pmatrix}. \tag{3.1.3}$$

Notice that matrix addition only makes sense if A and B are *both* $m \times n$, which makes sense, since I established that only vectors living in the same $\mathbb{R}^n$ may be added together consistently.

Example: We cannot add the matrices

$$A = \begin{pmatrix} -1 & 3 & 5 \\ 0 & 1 & -1 \end{pmatrix}, \quad B = \begin{pmatrix} 1 & 0 \\ 0 & -2 \end{pmatrix} \tag{3.1.4}$$

together since A is 2×3 and B is 2×2.

Example: It is possible to compute $A + 2B$ where

$$A = \begin{pmatrix} -1 & 3 & 5 \\ 0 & 1 & -1 \end{pmatrix}, \quad B = \begin{pmatrix} 1 & 0 & -2 \\ 0 & 6 & 2 \end{pmatrix} \tag{3.1.5}$$

since the matrices have the same size. In fact,

$$A + 2B = \begin{pmatrix} -1 & 3 & 5 \\ 0 & 1 & -1 \end{pmatrix} + \begin{pmatrix} 2 & 0 & -4 \\ 0 & 12 & 4 \end{pmatrix} = \begin{pmatrix} 1 & 3 & 1 \\ 0 & 13 & 3 \end{pmatrix}. \tag{3.1.6}$$

Before I move on, let me introduce some special terminology for matrices, which we'll use again and again in this course.

Definition: The **diagonal entries** of a matrix are the components a_{11} , a_{22} , ... In other words, the diagonal entries are the components of the matrix beginning with the upper left hand entry and moving down and right.

For the matrices A and B in the previous example, the diagonal entries are -1 and 1 for A and 1 and 6 for B.

Definition: A **diagonal matrix** is a **square matrix** (i.e. an $n \times n$ matrix) where only the diagonal entries are nonzero.

For instance, both $\begin{pmatrix} 1 & 0 & 0 \\ 0 & 2 & 0 \\ 0 & 0 & 3 \end{pmatrix}$ and $\begin{pmatrix} 1 & 0 & 0 \\ 0 & 0 & 0 \\ 0 & 0 & 2 \end{pmatrix}$ are 3×3 diagonal matrices. Notice that, in the second case, one of the diagonal entries is 0. That doesn't contradict the definition of a diagonal matrix; I only require that the square matrix have 0 entries off the main diagonal and make no restriction on what real numbers can appear on the diagonal.

Example: The identity matrix $I_{n \times n}$ is a classic example of a diagonal matrix. The $n \times n$ identity matrix has all 1's along the main diagonal and has 0's everywhere else. Sometimes the identity is just written as I if it is clear from context that the matrix is $n \times n$ or as $\mathbf{1}$. The notation $\mathbf{1}$ is highly suggestive: as I'll show you, the identity matrix behaves very much like the number 1 in the sense that any matrix A times the identity (of the right size) is A. You can already show this for yourself in the case of vectors given the definition of matrix-vector product.

Definition: The **zero matrix** is an $m \times n$ matrix with all entries equal to $\mathbf{0}$. The zero matrix has the (unique) property $A + \mathbf{0} = \mathbf{0} + A = A$. Again, you are supposed to be able to tell the size of $\mathbf{0}$ from context. Sometimes, authors will be explicit about the size of the zero matrix by writing $\mathbf{0}_{m \times n}$ or just $0_{m \times n}$.

It is not too hard to show that the set of all $m \times n$ matrices is a vector space itself! In other words, it satisfies all the axioms defining an abstract vector space with the definitions of matrix addition and scalar multiplication. I'll leave the details to you: it amounts to checking that all of the axioms of a vector space are satisfied by the set of all $m \times n$ matrices.

3.2 The Matrix Product

We see that adding matrices and multiplying them by scalars is easy! Since I'm adding mathematical structure to the set of all matrices, you may wonder, at this point, if it makes sense to **multiply** matrices together. We do know how to multiply an $m \times n$ matrix on the right by an $n \times 1$ vector, so this could be a special case of a more general operation like vector addition and scalar multiplication.

The most immediate and useful definition of the **matrix product** for our purposes arises by thinking of matrices as linear transformations from $\mathbb{R}^n$ to $\mathbb{R}^m$. Remember, given an $m \times n$ matrix A, we can define $T(\vec{x}) = A\vec{x}$, $T : \mathbb{R}^n \to \mathbb{R}^m$, which is linear. Take a step back to precalculus, and remember how we put two functions f and g together. You may remember the terminology **function composition** written with the $\circ$ symbol.

The general idea is that we have two functions $f : D_f \to C_f$ and $g : D_g \to C_g$ where D_f, D_g are the domains of f and g, respectively, and C_f, C_g are the respective codomains. If it happens that the range of f (which is a subset of C_f) is also a subset of the domain of g, then every image $f(d)$ of f can be plugged directly into g to get an element of C_g, namely $g(f(d))$. This process gives us a new function from D_f directly to C_g called the composition of g with f:

Definition: With the conditions satisfied above, $(g \circ f)(d) = g(f(d))$, $d \in D_f$ is a well-defined function from D_f to C_g.

Example: Suppose $f(x) = 2x - 1$ and $g(x) = x^2$. Clearly, $f, g : \mathbb{R} \to \mathbb{R}$ and the range of f is $\mathbb{R}$, the set of all real numbers, as is the domain for g. The function composition of g with f exists and is equal to

$$(g \circ f)(x) = g(f(x)) = g(2x - 1) = (2x - 1)^2. \tag{3.2.1}$$

Notice also that the range of g (the set of all nonnegative reals or the interval $[0, \infty)$ if you like) is obviously a subset of the domain of f. So, it is possible to define

$$(f \circ g)(x) = f(g(x)) = f(x^2) = 2x^2 - 1. \tag{3.2.2}$$

As this example shows, the compositions $g \circ f$ and $f \circ g$ do not have to be the same function.

If we think about matrices as linear transformations (which are really just functions), then we can compose them when the sizes of the matrices are just right: assume A is $m \times n$ and

B is $n \times p$ and define

$$T(\vec{x}) = A\vec{x} \ , \ T : \mathbb{R}^n \to \mathbb{R}^m$$
$$S(\vec{x}) = B\vec{x} \ , \ S : \mathbb{R}^p \to \mathbb{R}^n. \qquad (3.2.3)$$

In words, S takes $\mathbb{R}^p$ to $\mathbb{R}^n$ and then T takes $\mathbb{R}^n$ to $\mathbb{R}^m$ so it makes sense to compose them to get a function from $\mathbb{R}^p$ to $\mathbb{R}^m$: $(T \circ S)(\vec{x}) = T(S(\vec{x})) = A(B\vec{x})$. OK, so what does this have to do with matrix multiplication? When we compose linear transformations, we get a linear transformation. This follows easily from the definitions. Linear transformations between Euclidean spaces are represented by matrices. So really,

$$(T \circ S)(\vec{x}) = A(B\vec{x}) = M\vec{x} \qquad (3.2.4)$$

for some matrix M, which is the standard matrix representation of $T \circ S$. As you might guess, M is the matrix product of A and B. In short, we *define* $A \cdot B = M$ so that $A(B\vec{x}) = (A \cdot B)\,\vec{x}$!

Definition: If A is $m \times n$ and B is $n \times p$, then the matrix product $A \cdot B$ is defined to be the standard matrix representation of the linear transformation $A(B\vec{x})$.

It is extremely important that the number of rows of B equals the number of columns of A for this definition to work. Also, the order $A \cdot B$ is important. I'll have more to say about this shortly. This is all well and good, but the discussion so far is very abstract. What are the components of $A \cdot B$? What is the size of $A \cdot B$?

The size is easy to determine: the linear transformation $A(B\vec{x})$ takes $\mathbb{R}^p \to \mathbb{R}^m$, therefore $A \cdot B$ must be an $m \times p$ matrix. So if you have two matrices A and B of size $m \times n$ and $n \times p$, then the size of $A \cdot B$ comes from dropping the n's to get $m \times p$footnote$m$ is still the first dimension and p is still the second. The order must be respected!

Now for the components. The standard matrix representation of a linear transformation $T : \mathbb{R}^p \to \mathbb{R}^m$ is $\big(T(\hat{e}_1), \ldots, T(\hat{e}_p)\big)$ where the $\hat{e}$'s are the standard basis vectors for $\mathbb{R}^p$. Hence, we need $A(B\hat{e}_i) = A \cdot \vec{b}_i$ to find the columns of the product $A \cdot B$. Here, $\vec{b}_j$ is the j-th column of B. Putting these vectors altogether into a matrix,

$$A \cdot B = \left(A\vec{b}_1, \ldots, A\vec{b}_p\right). \tag{3.2.5}$$

In summary, to compute $A \cdot B$, compute each of the matrix-vector products $A\vec{b}_i$, and put them all (in order) in a matrix. Easy!

Example: Multiply, if possible, $A \cdot B$ and $B \cdot A$ given

$$A = \begin{pmatrix} 1 & -3 & -3 \\ 0 & 1 & -2 \end{pmatrix}, \quad B = \begin{pmatrix} 1 & 1 \\ 0 & 2 \end{pmatrix}. \tag{3.2.6}$$

It turns out that $A \cdot B$ cannot be done. A is a 2×3 matrix, whereas B is a 2×2 matrix. The number of columns of A does not match the number of rows of B. But it is true that the number of columns of B matches the number of rows of A, so $B \cdot A$ can be computed. Define

$$\vec{a}_1 = \begin{pmatrix} 1 \\ 0 \end{pmatrix}, \quad \vec{a}_2 = \begin{pmatrix} -3 \\ 1 \end{pmatrix}, \quad \vec{a}_3 = \begin{pmatrix} -3 \\ -2 \end{pmatrix}, \tag{3.2.7}$$

which are the columns of A. Then

$$B \cdot A = (B\vec{a}_1, B\vec{a}_2, B\vec{a}_3) = \begin{pmatrix} 1 & -2 & -5 \\ 0 & 2 & -4 \end{pmatrix}. \tag{3.2.8}$$

Notice that we end up with a 2×3 matrix, which is exactly what you would expect from multiplying a 2×2 matrix and a 2×3 matrix in that order.

Example: Evaluate $A \cdot B$ and $B \cdot A$ for

$$A = \begin{pmatrix} 5 & 1 \\ -3 & 1 \end{pmatrix}, \quad B = \begin{pmatrix} 2 & 0 \\ -4 & -1 \end{pmatrix} \tag{3.2.9}$$

For this example, both $A \cdot B$ and $B \cdot A$ are defined. In fact, using the definition of matrix product

$$A \cdot B = \begin{pmatrix} 6 & -1 \\ -10 & -1 \end{pmatrix}, \quad B \cdot A = \begin{pmatrix} 10 & 2 \\ -17 & -5 \end{pmatrix} \tag{3.2.10}$$

Notice that $A \cdot B \neq B \cdot A$. This example demonstrates that matrix multiplication is **non-commutative** whereas multiplying real numbers a and b together is **commutative**; i.e. $ab = ba$. The order of the matrices in the product matters!

Example: Multiply $A \cdot B$ and $A \cdot C$ given

$$A = \begin{pmatrix} 3 & -12 \\ -1 & 4 \end{pmatrix}, \quad B = \begin{pmatrix} 5 & 9 \\ 3 & 4 \end{pmatrix}, \quad C = \begin{pmatrix} 1 & -3 \\ 2 & 1 \end{pmatrix}. \tag{3.2.11}$$

Using the definition, we see that both products are defined and

$$A \cdot B = A \cdot C = \begin{pmatrix} -21 & -21 \\ 7 & 7 \end{pmatrix}. \tag{3.2.12}$$

In this case, we have $A \cdot B = A \cdot C$ where $A \neq \mathbf{0}$. If a, b, and c are real numbers with $a \neq 0$, then $ab = ac$ would mean that $b = c$. This is known as the cancellation property of real numbers. Clearly, this is not the case for matrix multiplication! Just because $A \cdot B = A \cdot C$ and $A \neq \mathbf{0}$, you cannot conclude that $B = C$.

Example: Finally, multiply $A \cdot B$ given

$$A = \begin{pmatrix} 3 & -12 \\ -2 & 8 \end{pmatrix}, \quad B = \begin{pmatrix} 4 & 12 \\ 1 & 3 \end{pmatrix}. \tag{3.2.13}$$

In this case, we see that the product is $\mathbf{0}$, even though we are multiplying nonzero matrices together! This is in stark contrast to real number multiplication where $ab = 0$ means $a = 0$ or $b = 0$.

In conclusion, the matrix product is quirky. The order of matrix multiplication matters, the cancellation property doesn't hold, and two nonzero matrices can multiply together to produce $\mathbf{0}$. Clearly, you need to be careful when using matrix products.

Before I move on to an application, let me mention that there is an alternative definition of the matrix product in terms of components. This definition can be used to calculate the components of $A \cdot B$ (when the product is defined) directly from the components of A and

B. Usually, the alternative I present below is used in proving statements and theorems in linear algebra, as I'll show below with an application from graph theory.

Suppose A is $m \times n$ and B is $n \times p$ so that $A \cdot B$ is defined. We may write

$$(A \cdot B)_{ij} = \sum_{l=1}^{n} A_{il} B_{lj} \tag{3.2.14}$$

where the summation symbol $\sum_{l=1}^{n}$ means we add all of the quantities to the right of the sigma ($\sum$) starting at $l = 1$ and stopping when we reach $l = n$. If you want a good understanding of how and why this expression works, you should go back over our examples and rework them using this formula.

Definition: If A is an $n \times n$ matrix and if k is a positive integer, then A^k denotes the product of k copies of A; i.e., $A^k = \underbrace{A \cdot A \cdot \cdots \cdot A}_{k \text{ times}}$. Define $A^0 = I$ so long as $A \neq 0$.

Matrix powers are useful in theory and applications. Later on, I'll show you how powers of a matrix can be used to define functions of matrices like $\exp A$, which are invaluable in solving linear systems of differential equations. In Chapter 1, we saw an application of matrix powers when Richard, the manager of a package delivery company, confronted the problem of finding the shortest and, hence, cheapest route from Boston to Berlin using Mercury Airways. In the example, I constructed the adjacency matrix for a graph representing a simplified, though similar, problem. Here is the matrix again:

$$A = \begin{pmatrix} 0 & 1 & 0 & 1 & 0 & 1 & 0 \\ 1 & 0 & 0 & 1 & 0 & 0 & 0 \\ 0 & 0 & 0 & 1 & 1 & 0 & 1 \\ 1 & 1 & 1 & 0 & 1 & 1 & 0 \\ 0 & 0 & 1 & 1 & 0 & 0 & 1 \\ 1 & 0 & 0 & 1 & 0 & 0 & 0 \\ 0 & 0 & 1 & 0 & 1 & 0 & 0 \end{pmatrix}. \tag{3.2.15}$$

I mentioned that the (i, j)-entry (the number in row i and column j) of A^k is precisely the number of k-walks connecting vertices i and j. The reasoning is relatively straightforward:

suppose we wish to count the number of $(k+1)$-walks from i to j on the graph. Here, k is fixed. If you think about it, any $(k+1)$-walk from i to j is made up of a 1-walk from i to a vertex connected directly to i by an edge and a k-walk from this intermediate vertex to j. Let $I = \{v_1, \ldots, v_l\}$ be the set of all such intermediate vertices. Then we have

$$\text{Number of } (k+1)\text{-walks from } i \text{ to } j = \sum_{v \in I} \text{Number of } k\text{-walks from } v \text{ to } j. \tag{3.2.16}$$

Here, $\sum_{v \in I}$ just means that you add up all of the quantities appear to the right of the sigma $(\sum)$ for every $v \in I$. I make an **inductive assumption** here: I'll assume that the components of A^k tell us the number of k-walks between any two vertices on the graph. Certainly this is true for $A^1 = A$ by definition of the adjacency matrix. If I can show that A^{k+1} tells us the number of $(k+1)$-walks assuming that A^k does the same for k-walks, then I'm finished with the proof by induction[1].

By my inductive assumption and the definition of the adjacency matrix, I can write

$$\text{Number of } (k+1)\text{-walks from } i \text{ to } j = \sum_{l=1}^{n} A_{il} \left(A^k\right)_{lj}. \tag{3.2.17}$$

However, $\sum_{l=1}^{n} A_{il} \left(A^k\right)_{lj}$ is precisely $\left(A^{k+1}\right)_{ij}$ given the definition of the matrix product in terms of components outlined above! So this completes the proof.

As an exercise for you, try verifying directly that

$$A^2 = \begin{pmatrix} 3 & 1 & 1 & 2 & 1 & 1 & 0 \\ 1 & 2 & 1 & 1 & 1 & 2 & 0 \\ 1 & 1 & 3 & 1 & 2 & 1 & 1 \\ 2 & 1 & 1 & 5 & 1 & 1 & 2 \\ 1 & 1 & 2 & 1 & 3 & 1 & 1 \\ 1 & 2 & 1 & 1 & 1 & 2 & 0 \\ 0 & 0 & 1 & 2 & 1 & 0 & 2 \end{pmatrix}, \quad A^3 = \begin{pmatrix} 4 & 5 & 3 & 7 & 3 & 5 & 2 \\ 5 & 2 & 2 & 7 & 2 & 2 & 2 \\ 3 & 2 & 4 & 8 & 5 & 2 & 5 \\ 7 & 7 & 8 & 6 & 8 & 7 & 2 \\ 3 & 2 & 5 & 8 & 4 & 2 & 5 \\ 5 & 2 & 2 & 7 & 2 & 2 & 2 \\ 2 & 2 & 5 & 2 & 5 & 2 & 2 \end{pmatrix}. \tag{3.2.18}$$

[1] Inductive proofs are like knocking over dominoes: if you line up all of the dominoes and knock over the first one, all others fall down as well. If you are unfamiliar with the concept, I suggest looking up "proof by induction" online to get a general idea of how it works.

We know from Chapter 1 that these matrices do properly catalogue the number of 2-walks and 3-walks, respectively, for the example graph. Finally, notice that adjacency matrices in general look the same over the main diagonal; *i.e* the upper right hand part of the matrix above the main diagonal is the mirror image of the lower left hand part below the main diagonal. The same is true for the powers of the adjacency matrix (as you should verify). Adjacency matrices have a nice **symmetry** over the main diagonal and are called **symmetric**.

Definition: An $n \times n$ matrix A is **symmetric** if and only if $A_{ij} = A_{ji}$ for all i and j. More compactly, A is symmetric if and only if $A^T = A$, where T indicates the transpose operation.

Symmetric matrices have many nice properties, some of which I'll touch on during our discussion of eigenvectors, eigenvalues, and inner product spaces.

3.3 The Matrix Inverse

Before I plunge into the topic of the matrix inverse, recall something very basic from algebra. For $a \in \mathbb{R}$ and $a \neq 0$, we know that we can solve $ax = b$ by dividing both sides by a, which is equivalent to multiplying by a^{-1}. Remember that for $a \neq 0$, $aa^{-1} = a^{-1}a = 1$. It is natural to wonder if a linear system like $A\vec{x} = \vec{b}$ may be solved in an analogous way. Putting it another way, is it possible to define A^{-1} that works in the same way for linear system as a^{-1} does for $ax = b$? Putting it yet another way, is it possible to define a A^{-1} so that $AA^{-1} = A^{-1}A = I$?

One way of answering this question is to once again think of a matrix A as defining a linear transformation. Suppose A is $m \times n$, and define $T(\vec{x}) = A\vec{x}$. If a matrix A^{-1} exists in the sense above, then $S(\vec{x}) = A^{-1}\vec{x}$ has to be the **inverse function** for T. Recall the definition of an inverse from precalculus:

Definition: Suppose $f : D_f \to C_f$, $g : D_g \to C_g$ are functions with the range of f equal to the domain of g. Then if $(g \circ f)(x) = x$ and $(f \circ g)(y) = y$ for all $x \in D_f$ and $y \in D_g$, then f and g are inverses of each other, and we write $f^{-1} = g$.

In a sense, the inverse of a function "undoes" the action of f. For instance, $f(x) = x^2$, $x \in [0, \infty)$ and $g(x) = \sqrt{x}$, $x \in [0, \infty)$ are inverses of each other. If A is $m \times n$, then $T : \mathbb{R}^n \to \mathbb{R}^m$ where $T(\vec{x}) = A\vec{x}$. Moreover, if A^{-1} exists so that T has an inverse function, it has to be the case that T is both one-to-one and onto. I will now argue that $n = m$; there is no other possibility.

First, I will show that T being one-to-one forces $m \geq n$. From the last chapter, we know that T is one-to-one if and only if the kernel of T is $\{\vec{0}\}$; i.e. the null space of A is the trivial space. This is equivalent to saying that the only solution to $A\vec{x} = \vec{0}$ is the trivial solution $\vec{x} = \vec{0}$. The linear system $A\vec{x} = \vec{0}$ is consistent and cannot possess free variables. Therefore, every column has a pivot since every variable is basic. Since there are exactly n columns and hence n pivots, it has to be the case that the number of rows with pivots (which equals n) is less than or equal to the total number of rows (which is m). Therefore, $n \leq m$ as I claimed.

Second, I will show that T being onto forces $m \leq n$. Recall that T is onto if and only if the range of T is all of $\mathbb{R}^m$. In other words, $A\vec{x} = \vec{b}$ always has at least one solution for $\vec{b} \in \mathbb{R}^m$. By definition of the matrix-vector product, this means that the columns of A must span $\mathbb{R}^m$; i.e. Col $A = \mathbb{R}^m$. But $\mathbb{R}^m$ is m-dimensional, which means any basis must have exactly m elements. This, in turn, means that A has to have at least m columns: if there are more columns, then some subset is linearly independent. If we have fewer than m columns, that means that a basis for $\mathbb{R}^m$ can be found with fewer than m elements, which is impossible! So, in conclusion, the number of columns n is at least as big as m; i.e. $n \geq m$.

So if T is one-to-one and onto, it must be the case that $m \geq n$ and $m \leq n$, which can only be true if $m = n$. This means that if A^{-1} exists in the sense I laid out above, then A has to be $n \times n$; only square matrices can have inverses! But this doesn't mean that every square matrix has an inverse. For instance, no inverse exists for the matrix $\begin{pmatrix} 1 & 0 \\ 0 & 0 \end{pmatrix}$, as I'll demonstrate shortly[2].

Definition: An $n \times n$ matrix A is **invertible** if there is an $n \times n$ matrix denoted by

[2]Think about the corresponding linear transformation. Is it one-to-one? Is it onto?

A^{-1} so that

$$AA^{-1} = I_{n \times n} \quad \text{and} \quad A^{-1}A = I_{n \times n}. \tag{3.3.1}$$

We also say that an invertible matrix is **nonsingular**. A matrix without an inverse is called **singular**.

For 2×2 matrices, there is a simple test to determine if an inverse exists: compute the **determinant** of the matrix. If the determinant is 0, the matrix is singular; if the determinant is nonzero, then the matrix is invertible.

Definition: Given a matrix $A = \begin{pmatrix} a & b \\ c & d \end{pmatrix}$, define the **determinant** as follows:

$$\det A = ad - bc.$$

Basically, the determinant is calculated in a "cross" pattern: multiply the diagonal terms, multiply the off diagonal terms, and take the difference.

Example: I can determine if the matrix $A = \begin{pmatrix} 1 & 2 \\ 3 & 4 \end{pmatrix}$ is invertible by computing the determinant: $\det A = 1 \cdot 4 - 2 \cdot 3 = -2$. So, yes, the inverse does exist.

Example: I can now show that the matrix $A = \begin{pmatrix} 1 & 0 \\ 0 & 0 \end{pmatrix}$ is singular as claimed by computing the determinant: $\det A = 1 \cdot 0 - 0 \cdot 0 = 0$.

Given that the determinant is nonzero, how do we calculate the inverse? There is a simple formula known as **Cramer's Rule** for the inverse in the 2×2 case:

Cramer's Rule: Let $A = \begin{pmatrix} a & b \\ c & d \end{pmatrix}$ and suppose $\det A \neq 0$. Then

$$A^{-1} = \frac{1}{\det A} \begin{pmatrix} d & -b \\ -c & a \end{pmatrix},$$

which can be verified explicitly by multiplying A by this matrix in either order.

Example: Going back to $A = \begin{pmatrix} 1 & 2 \\ 3 & 4 \end{pmatrix}$, the matrix inverse is

$$A^{-1} = \begin{pmatrix} -2 & 1 \\ \frac{3}{2} & -\frac{1}{2} \end{pmatrix}. \tag{3.3.2}$$

I'll have to wait until the next chapter to define determinants for larger matrices. However, there is another way to test for invertibility of a general $n \times n$ matrix A that also provides us with an explicit expression for A^{-1} when it exists. The key is the row reduction algorithm!

To see how this works, remember that $A^{-1}\vec{x}$ maps $\mathbb{R}^n$ to $\mathbb{R}^n$ linearly. The matrix representation is of course A^{-1}, so we can compute the inverse by computing $A^{-1}\hat{e}_j$ for each of the standard basis vectors. Computing $A^{-1}\hat{e}_j$ is equivalent to finding the unique solution to $A\vec{x}_j = \hat{e}_j$, which can be accomplished by row reduction. But there is no need to solve each equation separately using row reduction; we can form the augmented matrix

$$(A \mid \hat{e}_1, \hat{e}_2, \ldots, \hat{e}_n) \tag{3.3.3}$$

and compute the rref to solve each equation $A\vec{x}_j = \hat{e}_j$ simultaneously.

If A^{-1} exists, then the left-hand part of the reduced augmented matrix must be equal to $I_{n \times n}$ after row reduction and the right-hand part is necessarily the matrix A^{-1} by construction. If the left hand side does not row reduce to $I_{n \times n}$, then you know immediately that A^{-1} cannot exist. So, you now have the following algorithm:

Algorithm for Finding the Inverse Matrix:

Row reduce the augmented matrix $(A \mid I)$. If A is row equivalent to I, then $(A \mid I)$ is row equivalent to $(I \mid A^{-1})$. Otherwise, A^{-1} does not exist.

Example: Use the algorithm to find the inverse of $\begin{pmatrix} 1 & 0 & 0 \\ 1 & 2 & 0 \\ 1 & 1 & 3 \end{pmatrix}$. First, I setup the augmented matrix $(A \mid I)$:

$$\left(\begin{array}{ccc|ccc} 1 & 0 & 0 & 1 & 0 & 0 \\ 1 & 2 & 0 & 0 & 1 & 0 \\ 1 & 1 & 3 & 0 & 0 & 1 \end{array}\right). \tag{3.3.4}$$

Using elementary row operations, it is possible to row reduce A to the identity. Using those very same row operations on the augmented matrix, I find

$$\left(\begin{array}{ccc|ccc} 1 & 0 & 0 & 1 & 0 & 0 \\ 1 & 2 & 0 & 0 & 1 & 0 \\ 1 & 1 & 3 & 0 & 0 & 1 \end{array}\right) \sim \left(\begin{array}{ccc|ccc} 1 & 0 & 0 & 1 & 0 & 0 \\ 0 & 1 & 0 & -\frac{1}{2} & \frac{1}{2} & 0 \\ 0 & 0 & 1 & -\frac{1}{6} & -\frac{1}{6} & \frac{1}{3} \end{array}\right) \tag{3.3.5}$$

from which I conclude that

$$A^{-1} = \begin{pmatrix} 1 & 0 & 0 \\ -\frac{1}{2} & \frac{1}{2} & 0 \\ -\frac{1}{6} & -\frac{1}{6} & \frac{1}{3} \end{pmatrix}. \tag{3.3.6}$$

To conclude this section, here are some useful facts about invertible matrices

1. If A is an invertible $n \times n$ matrix, then A^{-1} is invertible and

$$\left(A^{-1}\right)^{-1} = A. \tag{3.3.7}$$

2. If A and B are $n \times n$ invertible matrices, then so is AB, and the inverse of AB is given by

$$(AB)^{-1} = B^{-1}A^{-1}. \tag{3.3.8}$$

3. If A is an invertible matrix, then so is A^T, and the inverse of A^T is the transpose of A^{-1}:

$$\left(A^T\right)^{-1} = \left(A^{-1}\right)^T. \tag{3.3.9}$$

All of these properties can be deduced from the definition of A^{-1}. You should attempt the proof as an exercise.

3.4 Chapter 3 Exercises

1. Given the matrices $A = \begin{pmatrix} -1 & 0 & 4 \\ -1 & 2 & -1 \end{pmatrix}$, $B = \begin{pmatrix} 3 & -2 & -1 \\ 0 & -1 & 1 \end{pmatrix}$, $C = \begin{pmatrix} 3 & 0 \\ 0 & 2 \end{pmatrix}$, compute $C\,(A + (-2)B)$ and $(A + (-2)B)\,C$. If a matrix product does not exist, state why.

2. Given the matrices $A = \begin{pmatrix} 5 & 0 & -4 & 7 \\ 0 & 3 & 0 & -1 \\ 4 & 4 & 3 & -1 \end{pmatrix}$, $B = \begin{pmatrix} -1 & 0 & 0 & 2 \\ 8 & 9 & -1 & 0 \\ 6 & 4 & -3 & 0 \end{pmatrix}$, $C = \begin{pmatrix} 3 & 0 & 2 \end{pmatrix}$, compute $C\,(A + B)$ and $(A + B)\,C$. If a matrix product does not exist, state why.

3. Given $A = \begin{pmatrix} 5 & -1 \\ 0 & 3 \end{pmatrix}$, compute A^2, A^3 and A^4 by hand. Use the determinant test to show that the matrix has an inverse, and compute this inverse. Compute $(A^{-1})^2$ and $(A^{-1})^3$. Verify for all matrices computed that the determinant does not vanish, hence every matrix computed is invertible.

4. For the matrices $\begin{pmatrix} 5 & 0 & -4 \\ 0 & 3 & -1 \\ 4 & 4 & 3 \end{pmatrix}$, $\begin{pmatrix} 1 & 0 & 0 \\ 0 & 3 & 0 \\ 4 & 8 & -2 \end{pmatrix}$, and $\begin{pmatrix} 7 & -1 & 2 & 3 \\ 0 & 2 & -5 & 10 \\ -1 & 8 & 0 & 9 \\ 0 & 1 & 1 & 0 \end{pmatrix}$, compute the inverse matrix, if it exists, or demonstrate that the inverse does not exist. Use the Inverse Matrix Algorithm described in the text.

5. The derivative $\frac{d}{dx}$ is a linear transformation of the vector space

$$W = \text{Span}\{\sin(x)\ ,\ \cos(x)\ ,\ \sinh(x)\ ,\ \cosh(x)\}$$

onto itself. Demonstrate that an inverse transformation for the derivative exists on this space by constructing a matrix representation of the derivative and using the Inverse Matrix Algorithm. What are the images of each of the basis vectors of W under the inverse transformation?

6. Assuming that A and B are $n \times n$ invertible matrices, prove that $(A^T)^{-1} = (A^{-1})^T$ and $(AB)^{-1} = B^{-1}A^{-1}$.

7. The elementary row operations are used throughout this book in a variety of settings. In particular, I showed you how to use them to find the unique rref of a matrix. It turns out that these elementary row operations can be represented as matrices themselves. Generally, if an elementary row operation is performed on an $m \times n$ matrix A, the result is a matrix that may be written as EA, where the $m \times m$ matrix E is obtained by performing the same row operation on the $m \times m$ identity $I_{m \times m}$.

   In this exercise, I will try to convince you that this is true by having you look at a few examples. First, I'll fix the matrix A to be $\begin{pmatrix} 5 & 0 & -4 & -1 \\ 0 & 3 & -1 & 7 \\ 4 & 4 & 3 & -2 \end{pmatrix}$.

   (a) (Replacement) Perform the row operation $r_1 \to r_1 + 2r_2$ to obtain the row equivalent matrix $\begin{pmatrix} 5 & 6 & -6 & 13 \\ 0 & 3 & -1 & 7 \\ 4 & 4 & 3 & -2 \end{pmatrix}$. Perform the same row operation on $I_{3\times 3}$ to get $E = \begin{pmatrix} 1 & 2 & 0 \\ 0 & 1 & 0 \\ 0 & 0 & 1 \end{pmatrix}$. Compute EA.

   (b) (Scaling) Perform the row operation $r_3 \to (-2)r_3$ to obtain the row equivalent matrix $\begin{pmatrix} 5 & 0 & -4 & -1 \\ 0 & 3 & -1 & 7 \\ -8 & -8 & -6 & 4 \end{pmatrix}$. Perform the same row operation on $I_{3\times 3}$ to get $E = \begin{pmatrix} 1 & 0 & 0 \\ 0 & 1 & 0 \\ 0 & 0 & -2 \end{pmatrix}$. Compute EA.

   (c) (Interchange) Perform the row operation $r_2 \leftrightarrow r_3$ to obtain the row equivalent matrix $\begin{pmatrix} 5 & 0 & -4 & -1 \\ 4 & 4 & -3 & -2 \\ 0 & 3 & -1 & 7 \end{pmatrix}$. Perform the same row operation on $I_{3\times 3}$ to get $E = \begin{pmatrix} 1 & 0 & 0 \\ 0 & 0 & 1 \\ 0 & 1 & 0 \end{pmatrix}$. Compute EA.

8. Argue that the **elementary matrices** representing elementary row operations (see

previous problem) are, in fact, invertible. One approach would be to argue that the inverse of E is the elementary matrix of the same type (i.e. replacement, interchange, scaling) that transforms E back into the identity.

9. A consequence of the Inverse Matrix Algorithm is that an $n \times n$ matrix A is invertible if and only if it is row equivalent to the identity matrix $I_{n \times n}$. Prove that if A is invertible, then it is possible to write A as $(E_1 \dots E_s) I_{n \times n}$ for a collection of elementary matrices $E_1, \dots, E_s$.

10. In this final exercise, I ask you to prove a theorem characterizing invertible matrices. This exercise is challenging and may be skipped, though you should remember this theorem if you are working with invertible matrices: it may be useful to refer back to it later.

    Let A be an $n \times n$ matrix. Then the following statements are either all true or all false.

    (a) A is an invertible matrix.

    (b) A is row equivalent to the $n \times n$ identity matrix.

    (c) A has n pivot positions.

    (d) The equation $A\vec{x} = \vec{0}$ has only the trivial solution.

    (e) The columns of A form a linearly independent set.

    (f) The linear transformation $\vec{x} \to A\vec{x}$ is one-to-one.

    (g) The equation $A\vec{x} = \vec{b}$ has at least one solution for each $\vec{b} \in \mathbb{R}^n$.

    (h) The columns of A span $\mathbb{R}^n$.

    (i) The linear transformation $\vec{x} \to A\vec{x}$ is onto.

    (j) There is an $n \times n$ matrix C such that $CA = I$.

    (k) There is an $n \times n$ matrix D such that $AD = I$.

    (l) A^T is invertible.

    As suggested in [5], you may want to proceed according to the route: $a \to k \to g \to a$, $g \leftrightarrow h \leftrightarrow i$, $d \leftrightarrow e \leftrightarrow f$, $a \leftrightarrow l$.[3]

[3] The statement of the theorem and the strategy for proving it appear here as they do in [5].

3.5 Chapter 3 References

If the reader wishes to explore the theory behind the theorems presented in this chapter or wishes to see more examples, I suggest consulting [5, 6]. In particular, the strategy for proving the inverse matrix theorem of exercise 10 is elaborated further in [5]. The reader wishing to know more about the principle of mathematical induction in particular and the style of mathematical proof in general should consult [8]. Another useful reference for mathematical induction is [9]. The proof for the claim that the n-th power of the adjacency matrix provides information on the number n-walks between vertices was largely inspired by [2].

Chapter 4

Determinants

In the last chapter, I defined the concept of the matrix inverse. For 2×2 matrices, I showed you a very useful tool that determines exactly when A^{-1} exists: the determinant. For the 2×2 matrix $\begin{pmatrix} a & b \\ c & d \end{pmatrix}$, the determinant is defined to be $\det A = ad - bc$ and A^{-1} exists precisely when $\det A \neq 0$. In this chapter, I will show you how determinants are defined for general $n \times n$ matrices. These determinants operate in precisely the same way as their 2×2 counterpart; they determine when the inverse of the $n \times n$ matrix exists. While this is certainly useful in of itself, the inverse of a matrix has limited utility compared to other concepts I have presented so far. It turns out that determining if A^{-1} exists or not is just the tip of the iceberg: determinants have many more applications besides determining invertibility. I'll show you some of these applications in this chapter.

4.1 The Recursive Definition of the Determinant

There is a relatively compact formula defining the determinant of a general $n \times n$ determinant. Before I state it (since it can be quite challenging to unpack and interpret), I will show you a few examples to build up to the general definition. Determinants are defined **recursively**. This means that the determinant of an $n \times n$ matrix is computed/defined in terms of determinants of $(n-1) \times (n-1)$ submatrices, which are, in turn, computed in terms of determinants of $(n-2) \times (n-2)$ submatrices and so on all the way down to a collection of 2×2 determinants. These determinants are computed using $\det A = ad - bc$. So, the definition is recursive in the sense that $\det A$ for A an $n \times n$ matrix relies on the definition for matrices of size $(n-1)\times(n-1)$.

Let's start with an example involving a 3×3 matrix. Define

$$A=\begin{pmatrix}-1 & 2 & 6\\ 0 & 1 & -1\\ 4 & 2 & 1\end{pmatrix}. \tag{4.1.1}$$

The procedure to compute the determinant of A goes like this: first, focus only on the first row of A consisting of the entries $a_{11}=-1$, $a_{12}=2$ and $a_{13}=6$. Start with the first entry of the first row (namely, -1). Now, forget about the first row and the first column; the matrix left over is 2×2:

$$A_{11}=\begin{pmatrix}1 & -1\\ 2 & 1\end{pmatrix}. \tag{4.1.2}$$

The subscript of A_{11} is *not* meant in this context to denote the $(1,1)$-entry of A; rather, here, it means the submatrix obtained from deleting row 1 and column 1 from A. Moving on to the second entry of row 1 (namely 2), it occupies the second column. And so the submatrix associated with this entry is

$$A_{12}=\begin{pmatrix}0 & -1\\ 4 & 1\end{pmatrix} \tag{4.1.3}$$

obtained by deleting row 1 and columns 2. Similarly, to the third entry of row 1 (namely 6) I associate the submatrix

$$A_{13}=\begin{pmatrix}0 & 1\\ 4 & 2\end{pmatrix}. \tag{4.1.4}$$

The submatrices I've constructed are all 2×2, which means that we know how to compute their determinants. The determinant of A is defined to be

$$\begin{aligned}\det A &= a_{11}\det A_{11} - a_{12}\det A_{12} + a_{13}\det A_{13}\\ &= (-1)\det\begin{pmatrix}1 & -1\\ 2 & 1\end{pmatrix} - 2\det\begin{pmatrix}0 & -1\\ 4 & 1\end{pmatrix} + 6\det\begin{pmatrix}0 & 1\\ 4 & 2\end{pmatrix}\\ &= -35.\end{aligned} \tag{4.1.5}$$

Note the appearance of the minus sign in front of the middle term. In general, the determinant "alternates" in sign: the definition singling out the first row given above begins with a "+" in front of $a_{11} \det A_{11}$ then has a "−" in front of $a_{12} \det A_{12}$ and so on, alternating until the number of columns is exhausted. From this calculation, I define the determinant of a 4×4 by way of an example:

Example: Define

$$A = \begin{pmatrix} -1 & 2 & -1 & 4 \\ 0 & 1 & -1 & -2 \\ 0 & 0 & 1 & -3 \\ 0 & 0 & 0 & 5 \end{pmatrix}. \tag{4.1.6}$$

If I mimic the procedure used above for the 3×3 example, I should focus on the first row consisting of $a_{11} = -1$, $a_{12} = 2$, $a_{13} = -1$, $a_{14} = 4$. The submatrices associated with the entries (obtained by deleting the first row and the column the entry occupies) are

$$A_{11} = \begin{pmatrix} 1 & -1 & -2 \\ 0 & 1 & -3 \\ 0 & 0 & 5 \end{pmatrix}, A_{12} = \begin{pmatrix} 0 & -1 & -2 \\ 0 & 1 & -3 \\ 0 & 0 & 5 \end{pmatrix}, A_{13} = \begin{pmatrix} 0 & 1 & -2 \\ 0 & 0 & -3 \\ 0 & 0 & 5 \end{pmatrix}, A_{14} = \begin{pmatrix} 0 & 1 & -1 \\ 0 & 0 & 1 \\ 0 & 0 & 0 \end{pmatrix} \tag{4.1.7}$$

with determinants

$$\det A_{11} = 5 \ , \ \det A_{12} = \det A_{13} = \det A_{14} = 0 \tag{4.1.8}$$

as you should verify using the last example as a template. The determinant of A is defined to be

$$\det A = a_{11} \det A_{11} - a_{12} \det A_{12} + a_{13} \det A_{13} - a_{14} \det A_{14} = -5. \tag{4.1.9}$$

4.2 Cofactors

To organize the general definition of the determinant as efficiently as possible, I will define **matrix cofactors** as follows:

Definition: Given a square matrix A, the (i, j) cofactor of A is $C_{ij} = (-1)^{i+j} \det A_{ij}$, where A_{ij} is the submatrix obtained by deleting the i-th row and the j-th column of A.

As I have defined the determinant in the case of $n \times n$ matrices for $n = 2, 3, 4$, I hope you see that $\det A = a_{11}C_{11} + \cdots + a_{1n}C_{1n}$ holds generally. This is called the **cofactor expansion across the first row**.

Definition: The **cofactor expansion across the *i*-th row** is

$$\det A = a_{i1}C_{i1} + \cdots + a_{in}C_{in},$$

whereas the **cofactor expansion down the *j*-th column** is

$$\det A = a_{1j}C_{1j} + \cdots + a_{nj}C_{nj}.$$

You may wonder what is so special about the first row in the definition of the determinant. The following theorem tells us that there is nothing special about the first row at all!

Theorem: The determinant of an $n \times n$ matrix A can be computed by a cofactor expansion across any row or down any column.

This theorem is extremely useful in that it allows us the freedom to choose the column or row to expand along: the determinant is the same no matter the choice. If you go back and look at the 4×4 example, it makes sense to cofactor expand along the first column or along the fourth row, since they have the most number of 0's. The more 0's in a column or row, the less work needed to compute the determinant, since the corresponding cofactors don't contribute to the total determinant.[1] In fact, the theorem above guarantees that the determinant of our 4×4 example is precisely the product of the diagonal entries.

For more complicated matrices, it is helpful to remember the following "matrix" of signs for computing cofactor expansions of determinants:

[1] Anything times 0 is 0.

Definition: The $\boldsymbol{n \times n}$ **matrix of signs** is defined to be

$$\begin{pmatrix} + & - & + & \cdots \\ - & + & - & \cdots \\ + & - & + & \cdots \\ \vdots & & & \end{pmatrix}.$$

From this matrix, you can work out the sign pattern necessary for any cofactor expansion:

Example: Cofactor expand the determinant of

$$A = \begin{pmatrix} 1 & 5 & 0 \\ 2 & 4 & -1 \\ 0 & -2 & -1 \end{pmatrix} \tag{4.2.1}$$

across the third row and down the third column. First, the 3×3 matrix of signs is $\begin{pmatrix} + & - & + \\ - & + & - \\ + & - & + \end{pmatrix}$. Expanding along the bottom row, we have

$$\det A = +0 \det \begin{pmatrix} 5 & 0 \\ 4 & -1 \end{pmatrix} - (-2) \det \begin{pmatrix} 1 & 0 \\ 2 & -1 \end{pmatrix} + (-1) \det \begin{pmatrix} 1 & 5 \\ 2 & 4 \end{pmatrix} = 4$$

where the alternating pattern $+, -, +$ comes from the third row of the matrix of signs. Expanding down the third column, we have

$$\det A = +0 \det \begin{pmatrix} 2 & 4 \\ 0 & -2 \end{pmatrix} - (-1) \det \begin{pmatrix} 1 & 5 \\ 0 & -2 \end{pmatrix} + (-1) \det \begin{pmatrix} 1 & 5 \\ 2 & 4 \end{pmatrix} = 4$$

where, again, the alternating pattern $+, -, +$ comes from the third column of the matrix of signs.

When I defined the determinant of a 4×4 matrix, I used a specific example where all of the nonzero entries were on or above the main diagonal. This is an example of a **triangluar matrix**: a square matrix where all of the nonzero entries are on and above (or on and below) the main diagonal. These matrices are particularly nice in that their determinants are easy to compute:

Theorem: If A is a triangular matrix, then $\det A$ is the product of the entries on the main diagonal of A.

This theorem is important for the following reason: once I show you how row operations affect a determinant, you may row reduce a given matrix into an echelon form and then use the theorem to compute the determinant. The proof is relatively simple: assume that A is upper triangular for simplicity. If I cofactor expand along the first column, the only contribution to the determinant is $a_{11} \det A_{11}$, since every entry below a_{11} is 0 by definition. The submatrix A_{11} is, in turn, upper triangular, so inductively, it has to be the case that $\det A = a_{11}a_{22} \ldots a_{nn}$ for any $n \times n$ triangular matrix.

Now, it is time to state how row operations affect the determinant of A:

Properties of Determinants under Row Operations:
Let A be a square matrix.

1. If a replacement operation turns A into B, then $\det B = \det A$.
2. A single interchange operation changes the sign of $\det A$.
3. If one row of A is multiplied by k to produce B, then $\det B = k \det A$.

Example: Let's use the above properties to compute the determinant of

$$A = \begin{pmatrix} 3 & 6 & 9 \\ -3 & -6 & -12 \\ 1 & 0 & 5 \end{pmatrix}$$

by row reducing to an echelon form of A and keeping track of the changes to $\det A$ due to the row operations. First, notice that the entries of the first and second rows have a common factor 3:

$$\det A = 3^2 \det \begin{pmatrix} 1 & 2 & 3 \\ -1 & -2 & -4 \\ 1 & 0 & 5 \end{pmatrix} \tag{4.2.2}$$

where I have taken out two factors of 3, one for each row. The replacement operation doesn't change the determinant, so

$$\det A = 3^2 \det \begin{pmatrix} 0 & 2 & -2 \\ 0 & 0 & -1 \\ 1 & 0 & 5 \end{pmatrix} \tag{4.2.3}$$

upon replacing row 2 with row 2 plus row 1 and row 1 with row 1 minus row 3. I can pull out the common factor 2 from row 1 and perform two interchange operations:

$$\det A = 3^2 \cdot 2 \det \begin{pmatrix} 1 & 0 & 5 \\ 0 & 1 & -1 \\ 0 & 0 & -1 \end{pmatrix}. \tag{4.2.4}$$

Since I interchanged rows twice, the sign remains the same because $(-1)^2 = 1$. The matrix in the determinant is upper triangular. So $\det A = 3^2 \cdot 2 \cdot 1 \cdot 1 \cdot (-1) = -18$. Of course, this isn't the only way to find the determinant using row operations. I challenge you to find an alternative route and show that the determinants are the same.

In addition to these properties of determinants, here are some others that will prove useful throughout this book:

Theorem:

1. A square matrix is invertible iff $\det A \neq 0$ [2].
2. If A is $n \times n$, then $\det A^T = \det A$.
3. If A and B are $n \times n$ matrices, $\det AB = \det A \det B$.

4.3 Cross Product

The dot product is just one way to combine two vectors. In the case of the dot product, the yield is a scalar. The dot product can be used to compute the lengths of vectors and to find the angle between two nontrivial vectors. The **cross product**, on the other hand, is a way

[2]This just formalizes the original motivation for introducing the determinant for $n \times n$ matrices.

of combining two vectors in $\mathbb{R}^3$ to find another *vector* in $\mathbb{R}^3$:

Definition: Let $\vec{v}$ and $\vec{w}$ be in $\mathbb{R}^3$. Let $\vec{v} = v_1\hat{e}_1 + v_2\hat{e}_2 + v_3\hat{e}_3$ and $\vec{w} = w_1\hat{e}_1 + w_2\hat{e}_2 + w_3\hat{e}_3$. Then $\vec{v} \times \vec{w} = \det \begin{pmatrix} \hat{e}_1 & \hat{e}_2 & \hat{e}_3 \\ v_1 & v_2 & v_3 \\ w_1 & w_2 & w_3 \end{pmatrix}$.

There are few funny features about this definition. First, the cross product definition only works (uniquely) in $\mathbb{R}^3$, because it is defined in terms of a determinant (which is defined only for $n \times n$ matrices), and takes in two vectors and returns a third. More importantly, the first row is made up of *vectors* and not numbers. This is fine, since I am merely writing, in a more compact way, the alternative definition

$$\vec{v} \times \vec{w} = \det \begin{pmatrix} v_2 & v_3 \\ w_2 & w_3 \end{pmatrix} \hat{e}_1 - \det \begin{pmatrix} v_1 & v_3 \\ w_1 & w_3 \end{pmatrix} \hat{e}_2 + \det \begin{pmatrix} v_1 & v_2 \\ w_1 & w_2 \end{pmatrix} \hat{e}_3, \tag{4.3.1}$$

which is a perfectly valid vector expression.

Example: I can compute the cross product, of $\vec{v} = \begin{pmatrix} 1 \\ 0 \\ -1 \end{pmatrix}$ and $\vec{w} = \begin{pmatrix} 0 \\ 2 \\ -1 \end{pmatrix}$ using this formula to obtain:

$$\vec{v} \times \vec{w} = \det \begin{pmatrix} 0 & -1 \\ 2 & -1 \end{pmatrix} \hat{e}_1 - \det \begin{pmatrix} 1 & -1 \\ 0 & -1 \end{pmatrix} \hat{e}_2 + \det \begin{pmatrix} 1 & 0 \\ 0 & 2 \end{pmatrix} \hat{e}_3 = \begin{pmatrix} 2 \\ 1 \\ 2 \end{pmatrix} \tag{4.3.2}$$

The most immediate application of the cross product (explaining the strange definition) is that it produces a third vector perpendicular to two given nontrivial and nonparallel vectors. The vector cross product also has several applications in physics. For instance, the force experienced by a charged particle in a magnetic field $\vec{B}$ is given by $\vec{F} = q\vec{v} \times \vec{B}$. This is known as the Lorentz force law. The cross product also defines the Poynting vector $\vec{S} \propto \vec{E} \times \vec{B}$, which contains information about the momentum of an electromagnetic field. [3] It is also true that

$$\|\vec{v} \times \vec{w}\| = \|\vec{v}\| \|\vec{w}\| \sin\theta, \tag{4.3.3}$$

[3] Here, $\vec{E}$ is the electric field, and $\vec{B}$ is the magnetic field.

which is similar to the formula I gave you involving the dot product, only there we had a cosine instead of a sine. It is not hard to show that the quantity on the right, $\|\vec{v}\|\|\vec{w}\|\sin\theta$ is the area of the parallelogram defined by $\vec{v}$ and $\vec{w}$. This is an incredibly useful application of the cross product, because it gives us a geometric interpretation of the determinant: if A is square and defines the linear transformation $T : \mathbb{R}^n \to \mathbb{R}^n$, then $\det A$ measures how much T distorts the size of a given subset S of $\mathbb{R}^n$.

I'll show you how this works for a 2×2 matrix called A. Define $T(\vec{x}) = A\vec{x}$, and assume for simplicity that $\det A \neq 0$. Then, $T(\hat{e}_1)$ and $T(\hat{e}_2)$ define two nontrivial vectors in $\mathbb{R}^2$ that, in turn, define a parallelogram. The vectors $\hat{e}_1$ and $\hat{e}_2$ define a square with area 1. The area of the parallelogram defined by $T(\hat{e}_1)$ and $T(\hat{e}_2)$ is $\|T(\hat{e}_1) \times T(\hat{e}_2)\|$ from above. If I write $A = \begin{pmatrix} a & b \\ c & d \end{pmatrix}$, then

$$T(\hat{e}_1) = \begin{pmatrix} a \\ c \end{pmatrix}, \quad T(\hat{e}_2) = \begin{pmatrix} b \\ d \end{pmatrix}, \tag{4.3.4}$$

so a simple calculation shows that $\|T(\hat{e}_1) \times T(\hat{e}_2)\| = |\det A|$. Therefore, T distorts the area of the unit square defined by $\hat{e}_1$ and $\hat{e}_2$ by a factor of $|\det A|$. A more advanced treatment using calculus of several variables can generalize this result to show that $T : \mathbb{R}^n \to \mathbb{R}^n$ distorts volumes by a factor of $|\det A|$ if $T(\vec{x}) = A\vec{x}$, and A is nonsingular.

4.4 Chapter 4 Exercises

1. For the matrices $\begin{pmatrix} 5 & 0 & -4 \\ 0 & 3 & -1 \\ 4 & 4 & 3 \end{pmatrix}$, $\begin{pmatrix} 1 & 0 & 0 \\ 0 & 3 & 0 \\ 4 & 8 & -2 \end{pmatrix}$, and $\begin{pmatrix} 7 & -1 & 2 & 3 \\ 0 & 2 & -5 & 10 \\ -1 & 8 & 0 & 9 \\ 0 & 1 & 1 & 0 \end{pmatrix}$, compute the determinant using any method described in the text. Which, if any, of these matrices are invertible?

2. Compute the determinant of $A = \begin{pmatrix} 2 & 1 & -2 & 6 \\ 0 & -3 & -2 & 1 \\ 0 & 1 & 3 & 0 \\ 0 & 3 & -4 & 0 \end{pmatrix}$ to determine if A^{-1} exists.

3. If $\vec{u} = \begin{pmatrix} 3 \\ 4 \\ -1 \end{pmatrix}$ $\vec{v} = \begin{pmatrix} 2 \\ -2 \\ 0 \end{pmatrix}$, compute the cross product $\vec{u} \times \vec{v}$, and verify that $\vec{u} \times \vec{v}$ is orthogonal to both $\vec{u}$ and $\vec{v}$.

4. In this problem, you'll use the cross product to determine the equation of a plane in $\mathbb{R}^3$. The plane of interest contains the points $P = (-1, 2, 1), Q = (3, 0, 0)$, and $R = (0, -1, 0)$. A plane in $\mathbb{R}^3$ may be described by the equation $a\,(x - x_0) + b(y - y_0) + c(z - z_0) = 0$ where I am using (x, y, z)-coordinates, and (x_0, y_0, z_0) corresponds to the coordinates of a point on the plane. The constants a, b, c together define a vector $\vec{n} = (a, b, c)^T$ that is orthogonal to the plane. Such a vector is said to be **normal** to the plane. Compute a normal vector $\vec{n}$ by computing the cross product of the displacement vectors $\vec{PQ} = \vec{Q} - \vec{P}$ and $\vec{PR} = \vec{R} - \vec{P}$, which are necessarily contained in the plane. Verify that the points P, Q, and R do indeed reside in the plane you construct by plugging their coordinates into your plane equation.

5. The determinant of an $n \times n$ matrix A may be defined by a general formula that does not rely on the recursive notion, as described above. First, a little background on **permutations.** A **permutation of degree** n is a sequence $(k_1, \ldots, k_n)$ of positive integers not exceeding n, with the property that no two of the numbers are the same. For example, $(1, 2, 3)$ and $(3, 2, 1)$ are permutations of degree 3. I will use the Greek letter σ to indicate a permutation. You can get to any degree n permutation σ by starting with $(1, 2, 3, \ldots, n)$ and interchanging pairs of integers at each step. There are many ways of doing so, typically, but it is a general fact that the number of such interchanges is either always even or always odd. If the number of pair-wise interchanges is even, σ is called **even**; otherwise, it is **odd.**

   Define sgn σ to be 1 if σ is even and -1 if σ is odd. Then $\det A$ can be written as

   $$\det A = \sum_{\sigma} \operatorname{sgn} \sigma \; A_{1\sigma_1} A_{2\sigma_2} \ldots A_{n\sigma_n}.$$

   Here, the sum is over all possible degree n permutations and σ_i is the i-th entry of the permutation. If you wish, you can take this to be the definition of the determinant; it is tricky to show the two definitions are, in fact, the same.

Using this formula for $\det A$, verify the properties of the determinant outlined in the text. In other words, prove that the replacement row operation does not change the determinant, interchanging any pair of rows leads to a change of sign of the determinant, and a common multiple of the entries of a single row may be brought outside the determinant.

6. In this problem, I will guide you through a proof that $\det(AB) = \det A \det B$, where A and B are both $n \times n$. First, let's look at the case where we have $\det(EA)$, where E is an elementary matrix. Remember that E is supposed to represent a row operation (replacement, scaling, or interchange). Using the properties of determinants stated in the text and proved in the previous problem, show that $\det(EA) = \det(E)\det(A)$ by direct calculation. That is, compute $\det(E)$ for each type of elementary matrix and show that $\det(E)\det(A)$ does indeed equal $\det(EA)$, where EA is the matrix obtained by performing the corresponding elementary row operation.

   Next, recall that any matrix M has a unique rref and that $M = (E_1 \ldots E_p) R$ for some collection of elementary matrices $E_1, \ldots, E_p$ and unique rref matrix R. Since M is square (i.e. $n \times n$), it follows that R is necessarily diagonal, with 1's and 0's along the main diagonal. If R is not the identity, then $\det R$ is necessarily 0, as you should verify. Given what you established above, this means that $\det M$ is also 0.

   Returning to A and B, you should see that if one of A or B is not invertible, AB cannot be invertible. (Refer to the theorem proved at the end of the exercises of Chapter 3.) Therefore, $\det(AB) = \det A \det B$ trivially since both sides are 0. If both A and B are nonsingular, then they are row equivalent to the $n \times n$ identity. It follows that they are both products of elementary matrices. Use this to show that $\det(AB) = \det A \det B$ also holds in this case.

7. Argue that $\det A = 0$ if and only if A is not invertible as follows: if A is nonsingular, A^{-1} exists and $AA^{-1} = I$. Use the identity $\det(AB) = \det A \det B$ to show that $\det A \neq 0$. Assume that $\det A \neq 0$. Use the fact that A is the product of a series of elementary matrices and a rref R together with the identity $\det(AB) = \det A \det B$ to complete the proof.

4.5 Chapter 4 References

If the reader wishes to explore the theory behind the theorems presented in this chapter or wishes to see more examples, I suggest consulting [5, 6]. In particular, the strategy for proving $\det(AB) = \det A \det B$ in the exercises, as well as the definition of the determinant in terms of permutations, is largely inspired by [6].

Chapter 5

Eigenvalues and Eigenvectors

I showed you an important application in which calculating matrix powers is critical. Specifically, if we want to know the number of k-walks in a given graph, then we need to compute A^k where A is the adjacency matrix. If you went through the exercise of computing A^2 and A^3 for the 7×7 example adjacency matrix, you know that the calculation, even for low k, can be tedious, though possible. So what? A computer can easily handle such tedious work. But there is another important reason for exploring matrix powers that I'll elaborate on in this chapter.

If I give you a square matrix A, would e^A or $\sin(A)$ make sense to you at this point? Probably not. We know how to evaluate the exponential function and the trigonometric functions for numbers. But what does it mean to have a matrix as an argument? It turns out that functions like e^x and $\sin(x)$ can be defined in terms of infinite degree polynomials [1]. For example,

$$e^x = 1 + x + \frac{1}{2}x^2 + \frac{1}{6}x^3 + \dots, \tag{5.0.1}$$

which means that if we can handle arbitrary matrix powers, then we can define e^A:

$$e^A = I + A + \frac{1}{2}A^2 + \frac{1}{6}A^3 + \dots. \tag{5.0.2}$$

[1] Technically, we have to be very careful about defining such things as "infinite degree polynomials." If you take Calculus, you'll learn about Taylor series, which is the proper setting for this discussion. I won't assume you know anything about Taylor series, however.

Of course, I have to help you make sense of this process: do we really have to compute an infinite number of matrix powers? Does this infinite process ever terminate? One way of making sense of e^A is by way of **eigenvalues** and **eigenvectors**, the topic of this chapter.[2]

5.1 A Simple Example and Change of Basis

I'll start with a simple example to introduce the general ideas. Suppose I give you the matrix $A = \begin{pmatrix} 1 & -2 \\ -2 & 1 \end{pmatrix}$ and ask you to compute A^n for general n. If we fix n to say 5, it is possible to compute A^5 in finite time, but it would be difficult to extrapolate to a general formula for undetermined n.

Suppose, instead, that I give you $B = \begin{pmatrix} 3 & 0 \\ 0 & -1 \end{pmatrix}$ and ask you to compute B^n for general n. If you start with $n = 2, 3, 4, \ldots$, you very quickly will see that

$$B^n = \begin{pmatrix} 3^n & 0 \\ 0 & (-1)^n \end{pmatrix}. \tag{5.1.1}$$

In fact, it is generally true for a **diagonal matrix** $M = \text{diag}(\lambda_1, \lambda_2, \ldots, \lambda_p)$ that $M^n = \text{diag}(\lambda_1^n, \lambda_2^n, \ldots, \lambda_p^n)$ for any positive integer n. Here, a bit of notation: $\text{diag}(\lambda_1, \lambda_2, \ldots, \lambda_p)$ simply means the $p \times p$ matrix with diagonal entries λ_1, λ_2, etc. starting with the upper left entry and working down to the lower right. All other entries are 0. The Greek letter λ (pronounced "lambda") will be used to denote eigenvalues in this book, so it's best to get used to it as soon as possible.

So, you learn a valuable lesson: computing matrix powers of diagonal matrices is very easy. Now go back to our original discussion around matrices A and B. Finding a general formula for A^n by brute force is difficult, but the formula for B^n is easy to find, since it is diagonal. It turns out that B is just a different representation for A in a different basis! If I can find the connection between A and B, then I have hope of finding a formula for A^n.

As it is stated, A is a matrix in the standard representation; i.e. if $T(\vec{x}) = A\vec{x}$, then

[2]Unless stated otherwise, assume any given matrix in this chapter is square.

$A = (T(\hat{e}_1), T(\hat{e}_2))$. I showed you in previous chapters the usefulness of changing the basis to greatly simplify calculations. Observe what happens to the matrix components when I switch to a different basis.

The transformation $T : \mathbb{R}^2 \to \mathbb{R}^2$ maps vectors to vectors. Vectors are geometric objects (arrows) independent of how we choose to describe them (in terms of coordinates). I can represent $\vec{x} \in \mathbb{R}^2$ as $\vec{x} = x_1\hat{e}_1 + x_2\hat{e}_2$ in terms of the standard basis or as $\vec{x} = c_1\vec{v}_1 + c_2\vec{v}_2$ in terms of a different (nonstandard) basis $\{\vec{v}_1, \vec{v}_2\}$.

Using the linearity of T,

$$\begin{aligned}
[T(\vec{x})]_{\mathcal{B}} &= [T(c_1\vec{v}_1 + c_2\vec{v}_2)]_{\mathcal{B}} \\
&= [T(c_1\vec{v}_1)]_{\mathcal{B}} + [T(c_2\vec{v}_2)]_{\mathcal{B}} \\
&= c_1[T(\vec{v}_1)]_{\mathcal{B}} + c_2[T(\vec{v}_2)]_{\mathcal{B}} \\
&= \underbrace{([T(\vec{v}_1)]_{\mathcal{B}}, [T(\vec{v}_2)]_{\mathcal{B}})}_{[A]_{\mathcal{B}}} \begin{pmatrix} c_1 \\ c_2 \end{pmatrix} \\
&= [A]_{\mathcal{B}}\ [\vec{x}]_{\mathcal{B}}. \qquad (5.1.2)
\end{aligned}$$

Here, I have defined $[A]_{\mathcal{B}} = ([T(\vec{v}_1)]_{\mathcal{B}}, [T(\vec{v}_2)]_{\mathcal{B}})$ to be the **matrix representation of A with respect to $\mathcal{B}$**, at least for the two-dimensional case. In general,

$$[A]_{\mathcal{B}} = ([T(\vec{v}_1)]_{\mathcal{B}}, [T(\vec{v}_2)]_{\mathcal{B}}, \ldots, [T(\vec{v}_p)]_{\mathcal{B}}) \qquad (5.1.3)$$

for general basis $\mathcal{B} = \{\vec{v}_1, \ldots, \vec{v}_p\}$. This is the matrix to use to define T if we use the coordinate representation $[\vec{x}]_{\mathcal{B}}$ for $\vec{x}$ instead of the standard representation[3].

It would be nice to know exactly how $[A]_{\mathcal{B}}$ is related to $[A]_{\mathcal{B}'}$ for two distinct bases $\mathcal{B}, \mathcal{B}'$. Let $\vec{v}_i$ be some vector in the basis $\mathcal{B}$: there is a unique way of writing $\vec{v}_i$ in terms of the basis vectors of $\mathcal{B}' = \{\vec{w}_1, \ldots, \vec{w}_p\}$:

$$\vec{v}_j = c_{1j}\vec{w}_1 + \cdots + c_{pj}\vec{w}_p. \qquad (5.1.4)$$

[3]Technically, I should be writing $[A]_{\{\vec{e}_1,\ldots,\vec{e}_n\}}$ for A and $[\vec{x}]_{\{\vec{e}_1,\ldots,\vec{e}_n\}}$ for $\vec{x}$, but you can see how clunky that can be, so I just drop the notation for simplicity when the basis is the standard basis.

Putting the numbers c_{ij} into a matrix called $C_{\mathcal{B}\to\mathcal{B}'}$ (i.e. $(C_{\mathcal{B}\to\mathcal{B}'})_{ij} = c_{ij}$), it is generally true that

$$[\vec{x}]_{\mathcal{B}'} = C_{\mathcal{B}\to\mathcal{B}'}[\vec{x}]_{\mathcal{B}}. \tag{5.1.5}$$

The notation $\mathcal{B} \to \mathcal{B}'$ indicates that we are going *from* the basis $\mathcal{B}$ *to* the basis $\mathcal{B}'$. Moreover, since the numbers c_{ij} connect the two bases, it must be the case that the inverse of $C_{\mathcal{B}\to\mathcal{B}'}$ exists and $C^{-1}_{\mathcal{B}\to\mathcal{B}'} = C_{\mathcal{B}'\to\mathcal{B}}$.

Given the two bases, it follows that

$$[A]_{\mathcal{B}}\ [\vec{x}]_{\mathcal{B}} = C_{\mathcal{B}'\to\mathcal{B}} \cdot ([A]_{\mathcal{B}'}\ [\vec{x}]_{\mathcal{B}'}) \tag{5.1.6}$$

if the image vector of T is to be independent of the way I choose to describe it. If I write this expression entirely in terms of $[\vec{x}]_{\mathcal{B}}$, you see that

$$[A]_{\mathcal{B}}\ [\vec{x}]_{\mathcal{B}} = C_{\mathcal{B}'\to\mathcal{B}} \cdot ([A]_{\mathcal{B}'}\ C_{\mathcal{B}\to\mathcal{B}'}[\vec{x}]_{\mathcal{B}}) \tag{5.1.7}$$

or

$$[A]_{\mathcal{B}}\ [\vec{x}]_{\mathcal{B}} = (C_{\mathcal{B}'\to\mathcal{B}}\ [A]_{\mathcal{B}'}\ C_{\mathcal{B}\to\mathcal{B}'})[\vec{x}]_{\mathcal{B}}. \tag{5.1.8}$$

From this calculation, you see that it is entirely reasonable to assert

$$[A]_{\mathcal{B}} = C_{\mathcal{B}'\to\mathcal{B}}\ [A]_{\mathcal{B}'}\ C_{\mathcal{B}\to\mathcal{B}'} = C_{\mathcal{B}'\to\mathcal{B}}\ [A]_{\mathcal{B}'}\ (C_{\mathcal{B}'\to\mathcal{B}})^{-1}. \tag{5.1.9}$$

Definition: When two matrices A and B are related by $A = SBS^{-1}$, we say that the matrices are **similar**. The matrix S provides a **similarity transformation**.

The terminology **similar** comes from the fact that two **similar** square matrices A and B will not have the same components generally, but they will share the same important properties: similar matrices will have the same determinant, the same **trace** (which is defined to be the sum of the diagonal entries), the same eigenvalues, etc. Think of siblings: siblings are distinct individuals that have important characteristics in common.

I will now show that A and B defined above are similar through a change of basis. Don't worry about finding the basis; I'll give it to you now and show you later how to find such a basis on your own. The new basis is $\mathcal{B}' = \left\{ \begin{pmatrix} -1 \\ 1 \end{pmatrix}, \begin{pmatrix} 1 \\ 1 \end{pmatrix} \right\}$, and the old basis is $\mathcal{B} = \left\{ \begin{pmatrix} 1 \\ 0 \end{pmatrix}, \begin{pmatrix} 0 \\ 1 \end{pmatrix} \right\}$. By definition, we have

$$C_{\mathcal{B}\to\mathcal{B}'} = \begin{pmatrix} -\frac{1}{2} & \frac{1}{2} \\ \frac{1}{2} & \frac{1}{2} \end{pmatrix} \tag{5.1.10}$$

since

$$\hat{e}_1 = \begin{pmatrix} 1 \\ 0 \end{pmatrix} = \underbrace{\left(-\frac{1}{2}\right)}_{c_{11}} \begin{pmatrix} -1 \\ 1 \end{pmatrix} + \underbrace{\left(\frac{1}{2}\right)}_{c_{21}} \begin{pmatrix} 1 \\ 1 \end{pmatrix} \tag{5.1.11}$$

and

$$\hat{e}_2 = \begin{pmatrix} 0 \\ 1 \end{pmatrix} = \underbrace{\left(\frac{1}{2}\right)}_{c_{12}} \begin{pmatrix} -1 \\ 1 \end{pmatrix} + \underbrace{\left(\frac{1}{2}\right)}_{c_{22}} \begin{pmatrix} 1 \\ 1 \end{pmatrix}. \tag{5.1.12}$$

Inverting the matrix $C_{\mathcal{B}\to\mathcal{B}'}$, we have

$$C_{\mathcal{B}'\to\mathcal{B}} = (C_{\mathcal{B}\to\mathcal{B}'})^{-1} = \begin{pmatrix} -1 & 1 \\ 1 & 1 \end{pmatrix}. \tag{5.1.13}$$

It is a straightforward exercise to show that

$$A = \begin{pmatrix} 1 & -2 \\ -2 & 1 \end{pmatrix} = \underbrace{\begin{pmatrix} -1 & 1 \\ 1 & 1 \end{pmatrix}}_{C_{\mathcal{B}'\to\mathcal{B}}} \underbrace{\begin{pmatrix} 3 & 0 \\ 0 & -1 \end{pmatrix}}_{B} \underbrace{\begin{pmatrix} -\frac{1}{2} & \frac{1}{2} \\ \frac{1}{2} & \frac{1}{2} \end{pmatrix}}_{(C_{\mathcal{B}'\to\mathcal{B}})^{-1}}. \tag{5.1.14}$$

Hopefully you see also that $B = [A]_{\mathcal{B}'}$.

In short, I have shown you that a change of basis (which I gave you) allows us to find a similarity transformation between A and a *diagonal* matrix B. If we want A^n, we can use the following theorem (which you should work through on your own):

Theorem: If A and B are square matrices so that $A = SBS^{-1}$, then $A^n = SB^nS^{-1}$.

To show this, just multiply SBS^{-1} together n times, and show that pairs of the form $S^{-1}S$ reduce to just the identity. From above, we then have

$$A^n = \begin{pmatrix} -\frac{1}{2} & \frac{1}{2} \\ \frac{1}{2} & \frac{1}{2} \end{pmatrix} \begin{pmatrix} 3^n & 0 \\ 0 & (-1)^n \end{pmatrix} \begin{pmatrix} -1 & 1 \\ 1 & 1 \end{pmatrix} = \begin{pmatrix} \frac{(-1)^n}{2} + \frac{3^n}{2} & \frac{(-1)^n}{2} - \frac{3^n}{2} \\ \frac{(-1)^n}{2} - \frac{3^n}{2} & \frac{(-1)^n}{2} + \frac{3^n}{2} \end{pmatrix}. \tag{5.1.15}$$

Try a few values of n on your own to confirm this result.

Since we have a general formula for A^n, I can now write down an expression for e^A:

$$\begin{aligned} e^A &= I + A + \frac{1}{2}A^2 + \frac{1}{6}A^3 + \ldots \\ &= \begin{pmatrix} -\frac{1}{2} & \frac{1}{2} \\ \frac{1}{2} & \frac{1}{2} \end{pmatrix} \begin{pmatrix} e^3 & 0 \\ 0 & e^{-1} \end{pmatrix} \begin{pmatrix} -1 & 1 \\ 1 & 1 \end{pmatrix} \\ &= \begin{pmatrix} \frac{1}{2}e^{-1} + \frac{1}{2}e^3 & \frac{1}{2}e^{-1} - \frac{1}{2}e^3 \\ \frac{1}{2}e^{-1} - \frac{1}{2}e^3 & \frac{1}{2}e^{-1} + \frac{1}{2}e^3 \end{pmatrix}. \end{aligned} \tag{5.1.16}$$

5.2 Eigenvalues and Eigenvectors Defined

At this stage, it seems almost magical that the basis $\mathcal{B}'$ can transform A into a diagonal and, hence, simpler matrix. The magic, so to speak, comes from the fact that vectors making up $\mathcal{B}'$ constitute special directions for A. In fact, if you calculate the matrix-vector product of A with each basis vector, you find

$$A\begin{pmatrix} -1 \\ 1 \end{pmatrix} = \begin{pmatrix} 1 & -2 \\ -2 & 1 \end{pmatrix} \begin{pmatrix} -1 \\ 1 \end{pmatrix} = \begin{pmatrix} -3 \\ 3 \end{pmatrix} = 3\begin{pmatrix} -1 \\ 1 \end{pmatrix} \tag{5.2.1}$$

and

$$A\begin{pmatrix} 1 \\ 1 \end{pmatrix} = \begin{pmatrix} 1 & -2 \\ -2 & 1 \end{pmatrix} \begin{pmatrix} 1 \\ 1 \end{pmatrix} = \begin{pmatrix} -1 \\ -1 \end{pmatrix} = (-1)\begin{pmatrix} 1 \\ 1 \end{pmatrix}. \tag{5.2.2}$$

Notice that the overall effect of multiplying the basis vectors of $\mathcal{B}'$ by A is to scale each individual vector by a constant. In this sense, the basis vectors define special directions for A. You may also notice that the scale factors are the numbers on the main diagonal of the transformed matrix B. This is not a coincidence as I'll show in just a moment.

Definition: A nonzero vector $\vec{x}$ is called an **eigenvector** of A if $A\vec{x} = \lambda\vec{x}$ for some scalar λ. The corresponding scalar λ is called the **eigenvalue**.

Using this parlance, you see that $\vec{v}_1 = \begin{pmatrix} -1 \\ 1 \end{pmatrix}$ is an eigenvector of A with eigenvalue 3, and $\vec{v}_2 = \begin{pmatrix} 1 \\ 1 \end{pmatrix}$ is an eigenvector of A with eigenvalue -1. I use "an" instead of "the" since if $\vec{x}$ is an eigenvector of some matrix A with eigenvalue λ, then so is $c\vec{x}$ for any $c \neq 0$, as you can easily check.

Hopefully, the above example gives you a sense of how eigenvalues and eigenvectors can be useful. Indeed, eigenvalues and eigenvectors are among the most important topics of linear algebra, because they tie together so many of the previous topics and are used in so many different applications. One application I'll touch on at the end of this chapter is the solution and analysis of systems of differential equations. Differential equations are frequently used to model systems changing with time. If you stop to think about it, this application is very broad in and of itself. But eigenvectors and eigenvalues also appear in statistics and in physics. For instance, all observable quantities in quantum theory are represented by Hermitian matrices[4] for which eigenvalues are the possible measurement outcomes. Eigenvalues and eigenvectors are also important for understanding phase transitions of material bodies. Eigenvectors of a certain matrix[5] also define the principal axes of motion for a rigid body as it rotates around its center of mass. The list goes on and on.

5.3 Computing Eigenvalues

Now, it is time to start thinking about how to compute eigenvalues and eigenvectors. After all, you won't simply be handed the eigenvalues and eigenvectors for a matrix A every time! I'll start with the eigenvalues. From the definition, we know that an eigenvalue possesses a corresponding eigenvector $\vec{x} \neq \vec{0}$ such that $A\vec{x} = \lambda\vec{x}$. But, this equation is equivalent to the vector equation $(A - \lambda I)\vec{x} = \vec{0}$ after bringing the $\lambda\vec{x}$ term to the left and using the fact that $\lambda\vec{x} = (\lambda I)\vec{x}$ by definition of the identity matrix. This means that the square matrix $A - \lambda I$

[4] A Hermitian matrix H satisfies the condition $(H^*)^T = H$. Here, the star indicates complex conjugation, so we allow the possibility of complex components.

[5] This matrix is called the moment of inertia tensor.

necessarily has a nontrivial null space, because the eigenvector $\vec{x}$ (whatever it actually is) must be included in the null space. A square matrix has a nontrivial null space precisely when it is singular or noninvertible. From the last chapter, we know that this in turn is true if and only if the determinant of the matrix vanishes. So, in conclusion, λ is an eigenvalue of A if and only if $\det(A - \lambda I) = 0$!

If you expand out $\det(A - \lambda I)$ for an $n \times n$ matrix A, you find a degree n polynomial with variable λ. This polynomial is called the **characteristic polynomial** for A.

Example: If we return to our original example $A = \begin{pmatrix} 1 & -2 \\ -2 & 1 \end{pmatrix}$, then

$$A - \lambda I = \begin{pmatrix} 1 & -2 \\ -2 & 1 \end{pmatrix} - \lambda \begin{pmatrix} 1 & 0 \\ 0 & 1 \end{pmatrix} = \begin{pmatrix} 1-\lambda & -2 \\ -2 & 1-\lambda \end{pmatrix}, \tag{5.3.1}$$

so the characteristic polynomial is

$$\det \begin{pmatrix} 1-\lambda & -2 \\ -2 & 1-\lambda \end{pmatrix} = (1-\lambda)^2 - (-2)^2 = \lambda^2 - 2\lambda - 3. \tag{5.3.2}$$

The roots of this equation are precisely the eigenvalues of A: $\lambda^2 - 2\lambda - 3 = (\lambda + 1)(\lambda - 3)$, so the eigenvalues are -1 and 3. These are precisely the diagonal entries of B, by the way!

For the general case of an $n \times n$ matrix A, you either have to be lucky and have a polynomial that is easily factorable, or you have to use some numerical scheme to approximate the roots of the characteristic polynomial and, hence, the eigenvalues. However, there is a fact from algebra that guarantees that the roots of the characteristic polynomial exist even if you cannot find them explicitly.

Fundamental Theorem of Algebra: Every nonzero, single-variable, degree n polynomial with complex coefficients has exactly n complex roots, counting multiplicity.

Here, multiplicity just means that you have to count how many times a root appears in the list of all roots. For example, the polynomial $p(x) = (x-1)^2(x+3)^5(x-2)^4$ is a degree

11 polynomial with roots $1, -3$, and 2. This, at first glance, seems to contradict the fundamental theorem of algebra: clearly the list $1, -3, 2$ falls short of the supposed 11 roots. But look at the exponents of each of the factors: the exponent on the $x-1$ term is 2, the exponent on the $x+3$ term is 5, and the exponent on the $x-2$ term is 4. In this case, $r_1 = 1$ has multiplicity 2, $r_2 = -3$ has multiplicity 5, and $r_3 = 2$ has multiplicity 4. Add up all the multiplicities: $2+5+4 = 11$, which matches the degree! In this sense, you need to count the multiplicity, or number of times each root is repeated, in order for the fundamental theorem to hold.

So we know that the characteristic polynomial of an $n \times n$ matrix will have n roots, but these roots could be complex! It is entirely possible to have complex roots arise in a real-world problem. This will certainly be the case when we use eigenvalues and eigenvectors to understand linear systems of differential equations. As a more down-to-earth example, suppose we want the general expression for A^n for $A = \begin{pmatrix} 0 & -1 \\ 1 & 0 \end{pmatrix}$, similar to the example opening this chapter.

Based on that example, you know that eigenvalues and eigenvectors are extremely useful tools. So I'll calculate the eigenvalues:

$$\det(A - \lambda I) = 0 \rightarrow \det \begin{pmatrix} -\lambda & -1 \\ 1 & -\lambda \end{pmatrix} = \lambda^2 + 1 = 0. \tag{5.3.3}$$

The solutions to this equation are $\pm i$ where i is defined to be $\sqrt{-1}$, a complex number. The presence of i may make you feel uncomfortable, but you shouldn't be afraid of complex numbers; they can actually make many calculations easier to do, as I'll show you in the next chapter when I discuss Fourier series. A pair of corresponding eigenvectors are

$$\begin{pmatrix} i \\ 1 \end{pmatrix} \ (\lambda = i)\ , \ \begin{pmatrix} -i \\ 1 \end{pmatrix} \ (\lambda = -i) \tag{5.3.4}$$

as you can verify by direct substitution into $A\vec{x} = \lambda\vec{x}$. Mimicking the example at the beginning of the chapter, you see that

$$A^n = \begin{pmatrix} i & -i \\ 1 & 1 \end{pmatrix} \begin{pmatrix} i^n & 0 \\ 0 & (-i)^n \end{pmatrix} \begin{pmatrix} -\frac{i}{2} & \frac{1}{2} \\ \frac{i}{2} & \frac{1}{2} \end{pmatrix} = \begin{pmatrix} \frac{(-i)^n}{2} + \frac{i^n}{2} & \frac{(-i)^{n+1}}{2} + \frac{i^{n+1}}{2} \\ \frac{(-i)^{n-1}}{2} + \frac{i^{n-1}}{2} & \frac{(-i)^n}{2} + \frac{i^n}{2} \end{pmatrix}. \tag{5.3.5}$$

As an exercise to acclimate yourself to using i, compute A^n for $n = 0, 1, 2, 3, 4$. You should find that

$$A^0 = I\ ,\ A^1 = A\ ,\ A^2 = \begin{pmatrix} -1 & 0 \\ 0 & -1 \end{pmatrix}\ ,\ A^3 = \begin{pmatrix} 0 & 1 \\ -1 & 0 \end{pmatrix}\ ,\ A^4 = I. \tag{5.3.6}$$

There are few things to note here. First, even though the general formula for A^n involves i, you see that for $n = 0, \ldots, 4$ that A^n only has *real* components, as it should based on what we know of the definition of matrix powers. Second, if you compute A^n for n higher than 4, the list of matrices simply repeats what's written above. This is tied to the third and final observation: the astute reader will recognize that the A defined above is really a representation of the rotation transformation in $\mathbb{R}^2$ about the origin by a counterclockwise angle of $\frac{\pi}{2}$. Taking successive positive powers of A geometrically means performing a series of rotations of a given vector $\vec{v}$ about the origin, each of a counterclockwise angle of $\frac{\pi}{2}$. For instance, if you start with $\hat{e}_1$ pointing in the positive x_1-direction, $A\hat{e}_1$ is $\hat{e}_2$ pointing in the positive x_2-direction, $A^2\hat{e}_1$ is $-\hat{e}_1$ pointing in the negative x_1-direction and $A^3\hat{e}_1$ is $-\hat{e}_2$ pointing in the negative x_2-direction. $A^4\hat{e}_1$ is, of course, just $\hat{e}_1$, so the overall effect is to do nothing to this basis vector.[6] So, it shouldn't be surprising that $A^4 = I$ and higher powers of A simply repeat the list above: after applying A four times to a vector, you have rotated it back to its original position.

You may have noticed that the eigenvalues for $A = \begin{pmatrix} 0 & -1 \\ 1 & 0 \end{pmatrix}$ form a conjugate pair. In other words, one is the complex conjugate of the other. This isn't a coincidence: in general, the coefficients of the characteristic polynomial for a matrix A with *real* components will also be real. It is a simple consequence of complex conjugation that if a polynomial with real coefficients has a complex root, then the complex conjugate of the root is also a root. In short, for real characteristic polynomials, roots (or eigenvalues) come in complex conjugate

[6] Alternatively, it means we rotate it 360°, which amounts to the same thing.

pairs.

5.4 Computing Eigenvectors

Eigenvalues of a square matrix A are computed by way of the characteristic polynomial $\det(A - \lambda I) = 0$. Once the eigenvalues are determined, the eigenvectors may be computed using the following observation: once again, writing the defining equation for an eigenvector as $(A - \lambda I)\vec{x} = \vec{0}$, you see that the set of eigenvectors corresponding to λ is precisely the null space of $A - \lambda I$. Computing null spaces is something you know how to do!

Return to our original example $A = \begin{pmatrix} 1 & -2 \\ -2 & 1 \end{pmatrix}$. We know that the eigenvalues are 3 and -1; compute the eigenvectors for $\lambda = 3$ by computing the null space for $A - 3I$:

$$A - 3I = \begin{pmatrix} -2 & -2 \\ -2 & -2 \end{pmatrix}. \tag{5.4.1}$$

The equation $(A - 3I)\vec{x} = \vec{0}$ leads to the augmented matrix

$$\left(\begin{array}{cc|c} -2 & -2 & 0 \\ -2 & -2 & 0 \end{array}\right). \tag{5.4.2}$$

After a few simple row operations, this matrix is row equivalent to

$$\left(\begin{array}{cc|c} 1 & 1 & 0 \\ 0 & 0 & 0 \end{array}\right). \tag{5.4.3}$$

So if $\vec{x} = \begin{pmatrix} x_1 \\ x_2 \end{pmatrix}$ is an eigenvector for $\lambda = 3$, then x_1 is basic, and x_2 is free:

$$\vec{x} = \begin{pmatrix} x_1 \\ x_2 \end{pmatrix} = x_2 \begin{pmatrix} -1 \\ 1 \end{pmatrix}, \quad x_2 \in \mathbb{R}. \tag{5.4.4}$$

For any nonzero choice of x_2, we find an eigenvector for $\lambda = 3$. It is convenient for many applications to say that the set of eigenvectors is the nonzero vectors in the span of $\begin{pmatrix} -1 \\ 1 \end{pmatrix}$.

The span of this vector is referred to as the **eigenspace corresponding to $\boldsymbol{\lambda} = \mathbf{3}$.**

Definition: For an eigenvalue λ of A, the **eigenspace corresponding to $\boldsymbol{\lambda}$**, denoted $\text{Ei}(\lambda)$, is $\text{Null}(A - \lambda I)$.

A similar calculation to the one above shows that

$$\text{Ei}(\lambda = -1) = \text{Span}\left\{\begin{pmatrix} 1 \\ 1 \end{pmatrix}\right\}, \tag{5.4.5}$$

as you can readily verify.

I'll do one more example to drive home all of the major points I've covered so far.

Example: Compute the eigenvalues and corresponding eigenspaces for the matrix $A = \begin{pmatrix} 1 & 2 & 2 \\ 2 & 1 & 2 \\ 2 & 2 & 1 \end{pmatrix}$. First, I need to setup and solve the corresponding characteristic equation $\det(A - \lambda I) = 0$:

$$\det(A - \lambda I) = \det\begin{pmatrix} 1-\lambda & 2 & 2 \\ 2 & 1-\lambda & 2 \\ 2 & 2 & 1-\lambda \end{pmatrix} = \lambda^3 - 3\lambda^2 - 9\lambda - 5 = 0. \tag{5.4.6}$$

It's not immediately obvious how to solve this equation, though it is possible to have a computer do it for us. But where would the fun be in that? Luckily for me, the coefficients are all integers, so I can find a list of potential rational roots by using the rational root test: if the characteristic polynomial has a rational solution $\frac{p}{q}$, where p and q have no common factors, then p must divide the last term -5 and q must divide the coefficient of λ^3, namely 1. So, the only possible rational roots are $\pm 5, \pm 1$. It turns out that 5 and -1 are real roots, so the rational root test works very well here. You can verify by polynomial long division that -1 has multiplicity 2, so this is a complete list of roots: there are no others.

Now that I have the eigenvalues, I can find the corresponding eigenspaces. I'll start with

$\lambda = 5$: the equation $(A - 5I)\,\vec{x} = \vec{0}$ leads me to the augmented system

$$\left(\begin{array}{ccc|c} -4 & 2 & 2 & 0 \\ 2 & -4 & 2 & 0 \\ 2 & 2 & -4 & 0 \end{array}\right), \tag{5.4.7}$$

which has rref

$$\left(\begin{array}{ccc|c} 1 & 0 & -1 & 0 \\ 0 & 1 & -1 & 0 \\ 0 & 0 & 0 & 0 \end{array}\right). \tag{5.4.8}$$

Therefore, x_3 is free, $x_1 = x_3$, and $x_2 = x_3$. It follows immediately that

$$\text{Ei}(\lambda = 5) = \text{Span}\left\{ \begin{pmatrix} 1 \\ 1 \\ 1 \end{pmatrix} \right\}. \tag{5.4.9}$$

Now I'll look at $\lambda = -1$: the equation $(A - (-1)I)\,\vec{x} = \vec{0}$ leads me to the augmented system

$$\left(\begin{array}{ccc|c} 2 & 2 & 2 & 0 \\ 2 & 2 & 2 & 0 \\ 2 & 2 & 2 & 0 \end{array}\right), \tag{5.4.10}$$

which has rref

$$\left(\begin{array}{ccc|c} 1 & 1 & 1 & 0 \\ 0 & 0 & 0 & 0 \\ 0 & 0 & 0 & 0 \end{array}\right). \tag{5.4.11}$$

Therefore, x_2 , x_3 are free and $x_1 = -x_2 - x_3$. It follows immediately that

$$\text{Ei}(\lambda = -1) = \text{Span}\left\{ \begin{pmatrix} -1 \\ 1 \\ 0 \end{pmatrix}, \begin{pmatrix} -1 \\ 0 \\ 1 \end{pmatrix} \right\}. \tag{5.4.12}$$

5.5 Diagonalizable Matrices

Notice that each eigenspace for the example at the end of the last section has a dimension equal to the multiplicity of the corresponding eigenvalue: dim Ei$(\lambda = 5) = 1$, dim Ei$(\lambda = -1) = 2$. This does not always happen! Try computing the eigenvalues and eigenvectors of $A = \begin{pmatrix} 1 & 0 \\ 1 & 1 \end{pmatrix}$. You should find that the eigenvalues are $\lambda = 1$ with multiplicity 2, but the eigenspace Ei$(\lambda = 1) = \text{Span}\left\{\begin{pmatrix} 0 \\ 1 \end{pmatrix}\right\}$ is one-dimensional.

However, when you find yourself with a matrix that has n linearly independent eigenvectors in total, then that the matrix is **diagonalizable**:

Definition: An $n \times n$ matrix A is **diagonalizable** if A has n linearly independent eigenvectors.

So, $\begin{pmatrix} 1 & 2 & 2 \\ 2 & 1 & 2 \\ 2 & 2 & 1 \end{pmatrix}$ is diagonalizable, whereas $\begin{pmatrix} 1 & 0 \\ 1 & 1 \end{pmatrix}$ is not. What does diagonalizability mean, though? Since A (if diagonalizable) has n linearly independent eigenvectors, then we are free to use these eigenvectors as a basis for $\mathbb{R}^n$. Let $\mathcal{B}$ be the standard basis for $\mathbb{R}^n$ and $\mathcal{B}' = \{\vec{v}_1 , \dots , \vec{v}_n\}$ be the basis made up of linearly independent eigenvectors.

From our previous work, we know that $C_{\mathcal{B}' \to \mathcal{B}}$, the change of basis matrix from basis $\mathcal{B}'$ to basis $\mathcal{B}$, is given by

$$C_{\mathcal{B}' \to \mathcal{B}} = (\vec{v}_1 , \dots , \vec{v}_n). \tag{5.5.1}$$

Just for concision's sake, let's call this matrix V. Then

$$[A]_{\mathcal{B}'} = C_{\mathcal{B} \to \mathcal{B}'}[A]_{\mathcal{B}} \left(C_{\mathcal{B} \to \mathcal{B}'}\right)^{-1} = V^{-1}AV. \tag{5.5.2}$$

By definition of the matrix product

$$AV = A\left(\vec{v}_1 , \dots , \vec{v}_n\right) = \left(A\vec{v}_1 , \dots , A\vec{v}_n\right) \tag{5.5.3}$$

and, since each of the vectors is an eigenvector of A,

$$AV = (\lambda_1\vec{v}_1 \ , \ \ldots \ , \ \lambda_n\vec{v}_n) = VD \tag{5.5.4}$$

where $D = \text{diag}(\lambda_1, \ldots, \lambda_n)$ is a diagonal matrix. Therefore, $[A]_{\mathcal{B}'} = V^{-1}VD = D$, so the matrix is diagonal in the basis of eigenvectors. That's why mathematicians refer to such matrices as diagonalizable!

To diagonalize a matrix means to find matrices V and D such that D is diagonal and $A = VDV^{-1}$, as above. If the matrix A has n *distinct* eigenvalues, then you know that A automatically has n linearly independent eigenvectors and, hence, is diagonalizable. The utility of diagonalizable matrices should hopefully be apparent from our discussion above, but if not, I'll explain again here: if A is diagonalizable so that $A = VDV^{-1}$ for $D = \text{diag}(\lambda_1, \ldots, \lambda_n)$, then for many "nice" functions like $f(x) = e^x$ and $f(x) = \sin(x)$, we can define

$$f(A) = V \ \text{diag}(f(\lambda_1), \ldots, f(\lambda_n))V^{-1}. \tag{5.5.5}$$

If $\{\vec{v}_1, \ldots, \vec{v}_n\}$ consists of n linearly independent eigenvectors of A, then $V = (\vec{v}_1, \ldots, \vec{v}_n)$ and V^{-1} can, in principle, be computed using the inverse matrix algorithm.

Application: Linear Systems of Differential Equations

Perhaps one of the most important applications of eigenvalues and eigenvectors is to solve some **linear systems of first order differential equations**. Basically, differential equations are used to model systems that evolve continuously with time. If you stop to think about it, there are many systems in nature and the real world that fall under that category. For example, if we have a particle moving under the influence of a force field $\vec{F}(\vec{x})$, then Newton's second law tells us that

$$m\frac{d^2\vec{x}}{dt^2} = \vec{F}(\vec{x}(t)), \tag{5.5.6}$$

where $\vec{x}$ represents the position vector of the particle in three-dimensional space as a function of time.

In theory, solving this differential equation would allow us to find the position of the particle at any later time, assuming we know its position and velocity at a given initial time. In a sense, differential equations give us the power to predict the future! Another example comes from predator-prey models, or Lotka-Volterra equations. Consider populations of predators (described mathematically by $x_1(t)$), and prey (described by $x_2(t)$), where only the predators eat the prey and not vice versa. The interaction between the two populations can be modeled by

$$\frac{dx_2}{dt} = ax_2 - b\, x_1x_2\ , \quad \frac{dx_1}{dt} = -cx_1 + d\, x_1x_2 \tag{5.5.7}$$

where a, b, c, d are positive constants. The x_1x_2 terms link the populations together. Without those terms present (i.e. if $d = b = 0$), then the predators would exponentially die off due to starvation, and the prey would exponentially grow, because nothing is culling their population.

Differential equations occupy a broad area of mathematics with a long and interesting history. With the tools you've learned so far in this class, I can show you how to solve so-called linear systems of differential equations. The above examples are nonlinear, but I'll show you how linear systems can be used to understand nonlinear systems near **equilibrium points**, which I'll define soon.

Definition: Let $\vec{x}(t) \in \mathbb{R}^n$ be a vector function in n-dimensional space whose components vary smoothly as functions of time.[7] A **first order linear system of differential equations** has the form

$$\frac{d\vec{x}}{dt} = A\vec{x}(t), \tag{5.5.8}$$

where A is a constant $n \times n$ matrix. In addition, knowledge of $\vec{x}(0)$ is required to uniquely specify $\boldsymbol{x}(t)$ for all time $t > 0$. The vector $\vec{x}_0 = \vec{x}(0)$ is called the **vector of initial conditions**.

The system $\frac{d\vec{x}}{dt} = A\vec{x}(t)$ is linear, since if $\vec{u}(t)$ and $\vec{v}(t)$ are two separate solutions, then $c_1\vec{u}(t) + c_2\vec{v}(t)$ is also a solution. In other words, linear combinations of solutions produce new solutions. The difficulty with solving such linear systems is that the rate of change of any given component is **coupled**, or dependent, on the other components of $\vec{x}(t)$, if A is not diagonal.

[7]This means that you can take as many derivatives of the component functions as you like.

We can make great progress solving the system of equations if A happens to be a diagonalizable. First, recall a basic fact from calculus:

$$\frac{dy}{dt} = \lambda y \to y(t) = y(0)e^{\lambda t}, \tag{5.5.9}$$

which is one way of defining the number e. The equation $\frac{dy}{dt} = \lambda y(t)$ is very similar to $\frac{d\vec{x}}{dt} = A\vec{x}(t)$; in fact, it *is* a linear system with just a single component $x_1(t) = y(t)$. Let's increase the difficulty level by assuming that A is an $n \times n$ diagonal matrix: by definition of the matrix-vector product and assuming that $A = \text{diag}(\lambda_1, \ldots, \lambda_n)$,

$$\frac{dx_1}{dt} = \lambda_1 x_1 \;, \;\ldots\;, \;\frac{dx_n}{dt} = \lambda_n x_n, \tag{5.5.10}$$

so

$$x_1(t) = x_1(0)e^{\lambda_1 t} \;, \;\ldots\;, \;x_n(t) = x_n(0)e^{\lambda_n t}, \tag{5.5.11}$$

since we essentially have n separate equations identical to $\frac{dy}{dt} = \lambda y(t)$. You won't always have a diagonal A in most of the applications of interest, but it is often the case that A is *diagonalizable.* If A happens to be diagonalizable, then it is possible to decouple the system of equations by writing $A = VDV^{-1}$. In fact, it is easy to show that the solution will be $\vec{x}(t) = e^{tA}\vec{x}(0)$ when A is diagonalizable, just like the solution to $\frac{dy}{dt} = \lambda y(t)$. First, I'll compute the derivative of the matrix e^{tA}, called the **fundamental matrix**, with respect to t:

$$\frac{d}{dt}e^{tA} = \frac{d}{dt}\left\{ V \begin{pmatrix} e^{\lambda_1 t} & 0 & \ldots & \\ 0 & e^{\lambda_2 t} & 0 & \ldots \\ \vdots & & & \end{pmatrix} V^{-1} \right\} = V \begin{pmatrix} \lambda_1 e^{\lambda_1 t} & 0 & \ldots & \\ 0 & \lambda_2 e^{\lambda_2 t} & 0 & \ldots \\ \vdots & & & \end{pmatrix} V^{-1} = Ae^{tA}. \tag{5.5.12}$$

So, $\frac{d}{dt}\left(e^{tA}\vec{x}(0)\right) = Ae^{tA}\vec{x}(0)$, which implies that $e^{tA}\vec{x}(0)$ is the unique solution with the correct initial conditions.

The case $n = 2$ is particularly important. Here, we have a visualization tool in the x_1x_2-plane, called the **phase plane**, allowing us to condense most of the relevant information of a system of linear equations in a nice, compact picture.

Definition: A plot in the x_1x_2-plane that shows a representative sample of trajectories for a given 2×2 system is called a **phase portrait**.

It should hopefully be clear from the above calculation that the nature of the eigenvalues and the corresponding eigenvectors determine the nature of the general solution of the system $\frac{d\vec{x}}{dt} = A\vec{x}$. If we assume that A is a real-valued 2×2 matrix, there are three possibilities for the eigenvalues of A:

1. Both eigenvalues are real and different from each other.

2. The eigenvalues occur in complex conjugate pairs.

3. There is a single eigenvalue with multiplicity 2.

This follows from the basic theory of quadratic polynomials, since the characteristic polynomial has degree 2 by assumption. According to the quadratic formula, the three cases listed above are the only possibilities depending on the sign of the discriminant. In the second and third cases, you must be careful to make sure your matrix A is, in fact, diagonalizable. Diagonalizability is guaranteed if you have two linearly independent eigenvectors, but this may not happen if the eigenvalue is repeated. Since this is a book on linear algebra and I don't want to stray too far off topic into the realm of differential equations theory, I'll ignore cases where the eigenvalue is repeated.

Example: Consider the system of linear differential equations defined by $A = \begin{pmatrix} 3 & -1 \\ 2 & 1 \end{pmatrix}$. It is possible to diagonalize this matrix as

$$A = \begin{pmatrix} 3 & -1 \\ 2 & 1 \end{pmatrix} = \underbrace{\begin{pmatrix} 1+i & 1-i \\ 2 & 2 \end{pmatrix}}_{V} \underbrace{\begin{pmatrix} 2+i & 0 \\ 0 & 2-i \end{pmatrix}}_{D} \underbrace{\begin{pmatrix} -\frac{i}{2} & \frac{i+1}{4} \\ \frac{i}{2} & \frac{-i+1}{4} \end{pmatrix}}_{V^{-1}}, \tag{5.5.13}$$

which yields the fundamental matrix

$$e^{tA} = \begin{pmatrix} e^{2t}\left(\cos(t) + \sin(t)\right) & -e^{2t}\sin(t) \\ 2e^{2t}\sin(t) & e^{2t}\left(\cos(t) - \sin(t)\right) \end{pmatrix}. \tag{5.5.14}$$

For any initial conditions $\vec{x}_0 \neq \vec{0}$, the trajectories spiral out, away from the origin, since $\vec{x}(t) = e^{tA}\vec{x}_0$ and the components of the fundamental matrix involve e^{2t}; as t becomes large, e^{2t} grows without bound. If I choose the origin $\vec{0}$ as my initial point, then the solution remains at $\vec{0}$ for all time, since $\vec{x}(t) = e^{tA}\vec{0} = \vec{0}$. Since the system remains constant (in fact, $\vec{0}$) for all time, mathematicians say that $\vec{0}$ is an **equilibrium point**.

I can use a computer to generate the trajectories $\vec{x}$ for a large collection of initial conditions. (See Fig. 5.1.) From the phase portrait, we learn something interesting: if I choose

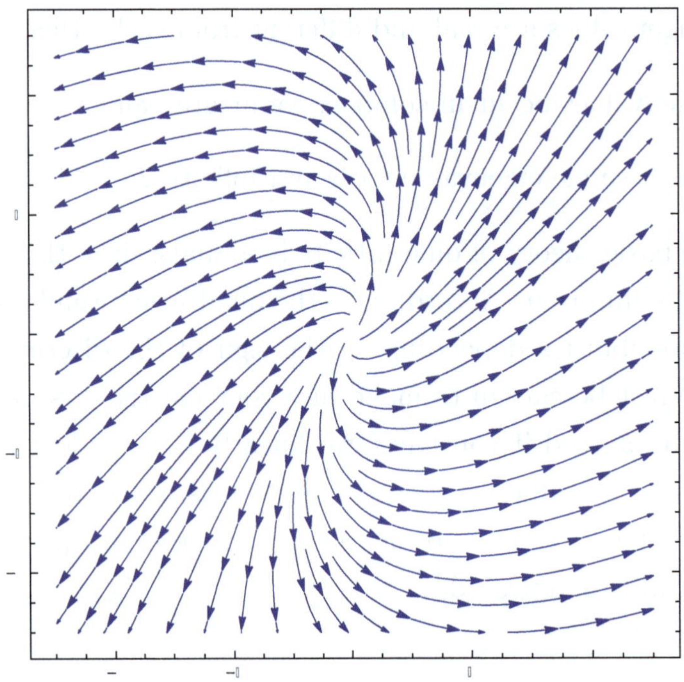

Figure 5.1: Phase portrait in phase space for the linear system $\frac{dx_1}{dt} = 3x_1 - x_2$, $\frac{dx_2}{dt} = 2x_1 + x_2$. Notice how trajectories tend away from the origin.

initial conditions near $\vec{0}$, the solution to the system of differential equations moves away from the origin exponentially. The equilibrium point here is **unstable**. Think of a pencil balanced on its tip. If you manage to balance it just right, it will stay upright for all time. But the slightest breeze or vibration will knock it down. This is the essence of an unstable equilibrium.

Example: As another example, consider the system defined by $A = \begin{pmatrix} -1 & 1 \\ -1 & -1 \end{pmatrix}$. This matrix may be diagonalized as

$$A = \begin{pmatrix} -1 & 1 \\ -1 & -1 \end{pmatrix} = \underbrace{\begin{pmatrix} -i & i \\ 1 & 1 \end{pmatrix}}_{V} \underbrace{\begin{pmatrix} -1+i & 0 \\ 0 & -1-i \end{pmatrix}}_{D} \underbrace{\begin{pmatrix} \frac{i}{2} & \frac{1}{2} \\ -\frac{i}{2} & \frac{1}{2} \end{pmatrix}}_{V^{-1}}, \tag{5.5.15}$$

as you should confirm by a direct calculation of the eigenvalues and eigenvectors. From this diagonalization of A, I compute the fundamental matrix to be

$$e^{tA} = \begin{pmatrix} e^{-t}\cos(t) & e^{-t}\sin(t) \\ -e^{-t}\sin(t) & e^{-t}\cos(t) \end{pmatrix}. \tag{5.5.16}$$

Just as in the previous example, I can use the fundamental matrix to generate a few trajectories $\vec{x}(t)$ to visualize the overall behavior of this system. (See Fig. 5.2.) The origin of the phase plane is an equilibrium point where the system remains unchanged for all time. In this case, however, the trajectories for initial values away from the origin tend towards the origin as t increases. This equilibrium is called **stable**. Think of a spring. If you push or pull on the spring, it will eventually tend back to its original length.

Example: As a final example, consider the system defined by $A = \begin{pmatrix} 4 & -5 \\ -2 & 1 \end{pmatrix}$. For this linear system, we will find that some trajectories approach the origin (again the equilibrium solution for this system) at first and then change direction and move away. The origin in this case is known as a **saddle point**. A diagonalization for A consists of

$$A = \begin{pmatrix} 4 & -5 \\ -2 & 1 \end{pmatrix} = \underbrace{\begin{pmatrix} -5 & 1 \\ 2 & 1 \end{pmatrix}}_{V} \underbrace{\begin{pmatrix} 6 & 0 \\ 0 & -1 \end{pmatrix}}_{D} \underbrace{\begin{pmatrix} -\frac{1}{7} & \frac{1}{7} \\ \frac{2}{7} & \frac{5}{7} \end{pmatrix}}_{V^{-1}}, \tag{5.5.17}$$

so the fundamental matrix is

$$e^{tA} = \begin{pmatrix} \frac{1}{7}e^{-t}\left(2+5e^{7t}\right) & -\frac{5}{7}e^{-t}\left(-1+e^{7t}\right) \\ -\frac{2}{7}e^{-t}\left(-1+e^{7t}\right) & \frac{1}{7}e^{-t}\left(5+2e^{7t}\right) \end{pmatrix}. \tag{5.5.18}$$

The phase portrait is depicted in Fig. 5.3. From these examples of 2×2 systems, you get

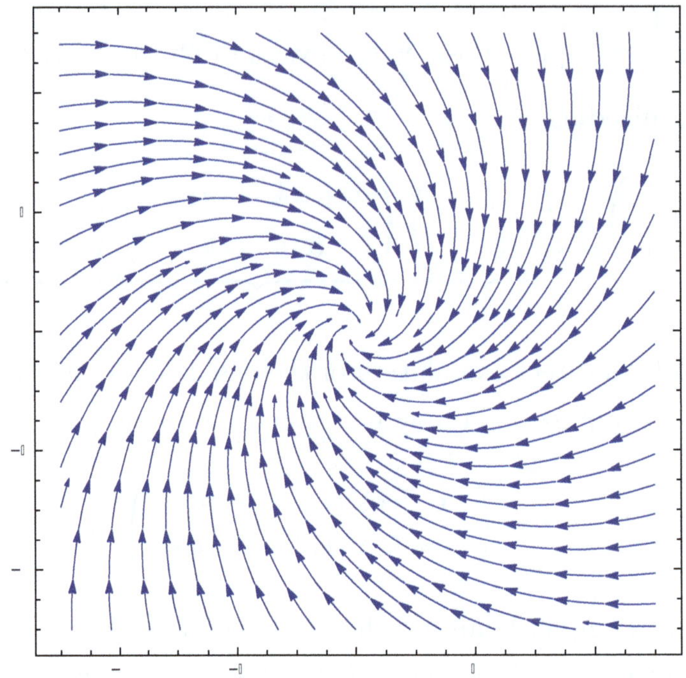

Figure 5.2: Phase portrait in phase space for the linear system $\frac{dx_1}{dt} = -x_1 + x_2$, $\frac{dx_2}{dt} = -x_1 - x_2$. Notice how trajectories tend towards the origin.

the sense that the behavior of the solutions near the equilibrium $\vec{0}$ is controlled by the eigenvalues of A: if the eigenvalues have nonnegative real parts, the solutions tend away from the origin whereas if they have negative real parts, the solutions tend towards the origin. When the eigenvalues are real with opposite sign, the behavior is more complicated, and we have a saddle at the origin. Of course, this is just the tip of the iceberg, and the student interested in learning more about differential equations should consult the references at the end of this chapter.

To conclude this section, I would like to give you a sense of how these linear systems of differential equations are used. This is by no means a complete list of examples, of course. In physics, for instance, linear systems of equations are used to solve order-n linear differential equations relevant to physical systems.

Example: Suppose a block is attached to a spring that moves in a viscous fluid in one direction. The friction of the fluid acts to slow the block down, while the spring supplies a

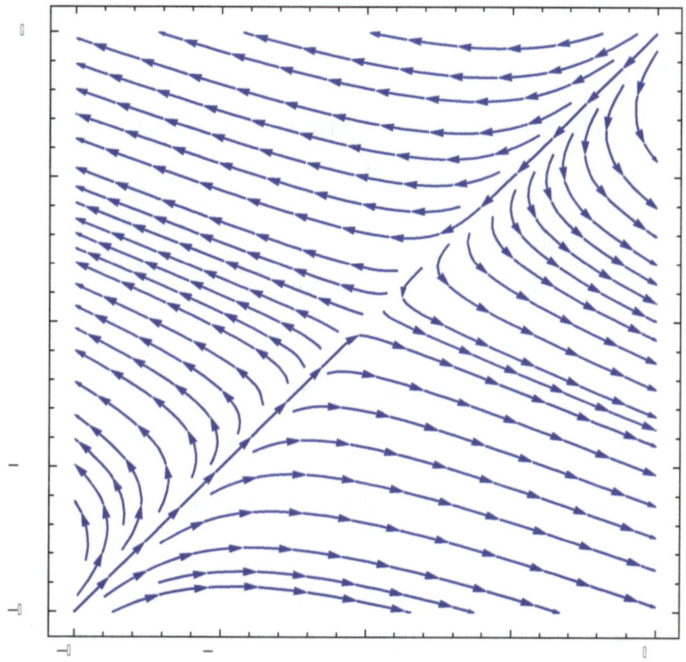

Figure 5.3: Phase portrait in phase space for the linear system $\frac{dx_1}{dt} = 4x_1 - 5x_2$, $\frac{dx_2}{dt} = -2x_1 + x_2$. Notice how some trajectories tend towards the origin and then move away.

linear restoring force. By a combination of Newton's second law and Hooke's spring law, we are led to the differential equation

$$m\frac{d^2x}{dt^2} = -kx - \gamma\frac{dx}{dt}, \tag{5.5.19}$$

where k is the spring constant (measuring the "stiffness" of the spring) and γ is the friction coefficient. It is possible to rewrite this single second order differential equation as a pair of first order linear differential equations by making the definitions $x_1(t) = x(t)$, $x_2(t) = \frac{dx}{dt}$:

$$\frac{dx_1}{dt} = x_2 \ , \ \frac{dx_2}{dt} = -\frac{k}{m}x_1 - \frac{\gamma}{m}x_2 \tag{5.5.20}$$

with corresponding coefficient matrix $A = \begin{pmatrix} 0 & 1 \\ -\frac{k}{m} & -\frac{\gamma}{m} \end{pmatrix}$.

Example: If a charged particle moves in a magnetic field $\vec{B}$ with velocity $\vec{v}$, then the Lorentz force law tells us that the force acting on the particle is $\vec{F} = q\vec{v} \times \vec{B}$, where q

is the charge on the particle and $\times$ is the vector cross product. Suppose, for simplicity, that the magnetic field is constant and points in the x_3 -direction; i.e. $\vec{B} = B_0\hat{e}_3$. Then $\vec{v} \times \vec{B} = -B_0 v_1 \hat{e}_2 + B_0 v_2 \hat{e}_1$. Newton's second law tells us that $m\frac{d\vec{v}}{dt} = \vec{F}$, so we have $\frac{dv_3}{dt} = 0$ (which means that the third component of the velocity is constant) and

$$m\frac{dv_1}{dt} = qB_0 v_2 \ , \ m\frac{dv_2}{dt} = -qB_0 v_1, \tag{5.5.21}$$

which is a linear system for the components of velocity with coefficient matrix $A = \begin{pmatrix} 0 & \frac{qB_0}{m} \\ -\frac{qB_0}{m} & 0 \end{pmatrix}$. As an exercise, you should solve this system and show that the motion of the particle is circular. Can you find the frequency at which the particle moves around in the magnetic field?

While linear systems appear in many contexts, nonlinear equations are far more common. However, linear systems of equations can be used to understand the behavior of a nonlinear system near an equilibrium point of a nonlinear system.

Example: Consider the predator-prey model

$$\frac{dx_2}{dt} = ax_2 - bx_1x_2 \ , \ \frac{dx_1}{dt} = -cx_1 + dx_1x_2 \tag{5.5.22}$$

where for concreteness, I will choose $a = c = 10$, $b = d = 1$. Granted, this may not be a terribly realistic model for an actual predator-prey system, but it illustrates the mathematical idea nicely. With this choice of constants, you should verify that $x_1 = 10 = x_2$ is an equilibrium solution of the system of differential equations. In other words, if you plug $x_1 = 10$ and $x_2 = 10$ into both sides of both equations, you'll find that both equations are true.

Now that we know that $(10, 10)$ is an equilibrium point, I want to determine whether or not the equilibrium is stable. To understand the behavior of solutions near this point, I assume solutions nearby have the form $x_1(t) = 10 + \delta x_1(t)$, $x_2(t) = 10 + \delta x_2(t)$[8] where δx_1 and δx_2 are "small" in the sense that powers of these functions like $\delta x_1(t)^2$ and products like $\delta x_1(t)\delta x_2(t)$ are negligible; i.e. they are so small that leaving them out entirely is a very

[8] δ is the Greek letter "delta."

good approximation. In technical parlance, I am **perturbing** the equilibrium.

If I plug these perturbations into the original predator-prey equations, ignoring powers and products of the delta terms, then the approximate system

$$\frac{d\delta x_1}{dt} \approx 10\delta x_2\ , \ \frac{d\delta x_2}{dt} \approx -10\delta x_1 \tag{5.5.23}$$

remains after the dust settles. This approximate linear system has the coefficient matrix $A = \begin{pmatrix} 0 & 10 \\ -10 & 0 \end{pmatrix}$ with eigenvalues $\lambda = \pm 10i$. The fundamental matrix $e^{tA} = \begin{pmatrix} \cos(10t) & \sin(10t) \\ -\sin(10t) & \cos(10t) \end{pmatrix}$ (which, incidentally, is a rotation matrix where the angle grows linearly with t) tells us that, if we perturb the system away from its equilibrium $(10, 10)$, the solutions tend to circle around the equilibrium or **orbit** it. The phase portrait for the full nonlinear system near the equilibrium $(10, 10)$ is shown in Fig. 5.4. In conclusion, useful information about the behavior of solutions to nonlinear systems near equilibrium points can be obtained by **linearizing** the system near the equilibrium point as I did in this example. However, linearization will not work for all nonlinear systems, unfortunately; see the exercises for more details.

5.6 Diagonalization of Symmetric Matrices

If you look back at the examples of the previous sections, quite a few of the diagonalizable matrices were **symmetric**. Remember, a square matrix A is symmetric if and only if $A = A^T$. It turns out that symmetric matrices have many nice properties when it comes to diagonalization. You may think that this is a highly specialized case and may not have much to do with the real world, but this is far from the truth! First, you know that adjacency matrices for graphs are symmetric matrices. Furthermore, symmetric matrices are used to define functions of vectors called **quadratic forms**, which are of the form $Q(\vec{x}) = \vec{x}^T A\vec{x}$. Quadratic forms come up frequently in optimization problems in engineering and in signal analysis. They are also used in physics, differential geometry, economics and statistics to name a few more examples. So, understanding symmetric matrices and their diagonalizations is far from a purely academic exercise.

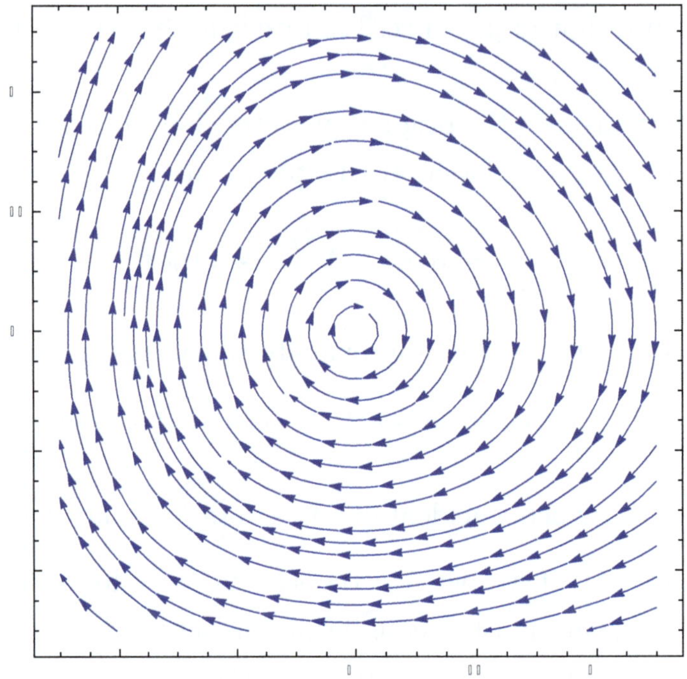

Figure 5.4: Phase portrait of the nonlinear system $\frac{dx_2}{dt} = 10x_2 - x_1x_2$, $\frac{dx_1}{dt} = -10x_1 + x_1x_2$ near the equilibrium point $(10, 10)$. Linearization predicts circular orbits near the equilibrium, which is reflected in the actual nonlinear phase portrait.

The first nice feature of symmetric matrices is that their eigenvalues are always real. This is surprising considering how many examples of the previous sections actually possessed complex eigenvalues. The proof of this statement requires us to define a new dot product for $\mathbb{C}^n$, the set of all n component vectors with *complex* entries. I will continue to use the overhead arrow notation for column vectors even when their components are complex just for clarity. I also have to generalize the notion of symmetry to complex matrices: I define the **Hermitian conjugate** of a complex matrix A to be $A^\dagger = (A^*)^T$, where the star means we have to take the complex conjugate of all the entries of A. Suppose $\vec{v}, \vec{w} \in \mathbb{C}^n$ and further suppose A is a **Hermitian matrix**, *i.e* $A^\dagger = A$. I define

$$\langle \vec{v}, \vec{w} \rangle = \vec{w}^\dagger \vec{v}, \tag{5.6.1}$$

which I hope you see reduces to the normal dot product $\vec{w} \cdot \vec{v}$ on $\mathbb{R}^n$ when the entries are all real. Then, in terms of the components of the vectors and a general complex matrix A,

$$\langle A\vec{v}, \vec{w}\rangle = \sum_{j,k=1}^{n} w_j^* A_{jk} v_k = \sum_{j,k=1}^{n} w_k^* A_{kj} v_j = \left(\sum_{j,k=1}^{n} w_k \left(A^*\right)_{kj} v_j^*\right)^* = \langle A^\dagger \vec{w}, \vec{v}\rangle^*. \tag{5.6.2}$$

Before I continue, let me point out a few important properties of the **inner product** $\langle \vec{v}, \vec{w}\rangle$. First, the inner product is distributive in each slot, i.e.

$$\langle \vec{v}_1 + \vec{v}_2, \vec{w}\rangle = \langle \vec{v}_1, \vec{w}\rangle + \langle \vec{v}_2, \vec{w}\rangle \ , \ \langle \vec{v}, \vec{w}_1 + \vec{w}_2\rangle = \langle \vec{v}, \vec{w}_1\rangle + \langle \vec{v}, \vec{w}_2\rangle. \tag{5.6.3}$$

Also, the complex conjugate operation switches the two vectors in their respective slots:

$$\langle \vec{v}, \vec{w}\rangle^* = \langle \vec{w}, \vec{v}\rangle. \tag{5.6.4}$$

Finally, for any complex number $c \in \mathbb{C}$,

$$\langle c\vec{v}, \vec{w}\rangle = c\langle \vec{v}, \vec{w}\rangle \ , \ \langle \vec{v}, c\vec{w}\rangle = c^*\langle \vec{v}, \vec{w}\rangle. \tag{5.6.5}$$

Notice that we have to be careful about the order of vectors in this new inner product, something we didn't have to worry about for the dot product of $\mathbb{R}^n$. Also, we need to be careful about pulling scalars out of the inner product; we may need to conjugate depending on where the scalar appears in the inner product.

From the above observations, you see that $\langle A\vec{v}, \vec{w}\rangle = \langle \vec{v}, A^\dagger \vec{w}\rangle$; i.e. we can move the matrix A between slots at the cost of a Hermitian conjugate. Suppose now that $\vec{v}$ and $\vec{w}$ are eigenvectors of the Hermitian matrix A with distinct eigenvalues λ_v and λ_w, respectively. I will show that not only must these eigenvalues be real, but also $\langle \vec{v}, \vec{w}\rangle = 0$.

First, $\langle A\vec{v}, \vec{v}\rangle = \langle \lambda_v \vec{v}, \vec{v}\rangle$, since $A\vec{v} = \lambda_v \vec{v}$. Furthermore, since the eigenvalue is in the first slot of the inner product, $\langle A\vec{v}, \vec{v}\rangle = \lambda_v \langle \vec{v}, \vec{v}\rangle$. Now, using the property established above, $\langle A\vec{v}, \vec{v}\rangle = \langle \vec{v}, A^\dagger \vec{v}\rangle$. Using the fact that $A^\dagger = A$, we have $\langle A\vec{v}, \vec{v}\rangle = \langle \vec{v}, A\vec{v}\rangle = \langle \vec{v}, \lambda_v \vec{v}\rangle$. Since λ_v appears in the *second* slot now, we have $\langle A\vec{v}, \vec{v}\rangle = \lambda_v^* \langle \vec{v}, \vec{v}\rangle$. So, the conclusion is that $\lambda_v^* \langle \vec{v}, \vec{v}\rangle = \lambda_v \langle \vec{v}, \vec{v}\rangle$. There are only two possibilities then: either $\lambda_v^* = \lambda_v$, implying that the eigenvalue is real, or $\langle \vec{v}, \vec{v}\rangle = 0$. In the latter case, this would mean that $\vec{v}^\dagger \vec{v} = 0$, which can only be true if $\vec{v} = \vec{0}$. But an eigenvector is necessarily nonzero, so we see that λ_v is real! The argument applies to all eigenvalues, so all eigenvalues of $A = A^\dagger$ must be real. Specifically, if A has only real entries so that $A = A^T$, the eigenvalues of symmetric matrices must also be real.

Consider, now, the inner product of $\langle A\vec{v}, \vec{w}\rangle$ where $\vec{v}$ and $\vec{w}$ correspond to distinct eigenvalues. Then,

$$\langle A\vec{v}, \vec{w}\rangle = \lambda_v\langle \vec{v}, \vec{w}\rangle = \langle \vec{v}, A\vec{w}\rangle = \lambda_w\langle \vec{v}, \vec{w}\rangle \tag{5.6.6}$$

where I have used the fact that $A = A^\dagger$ and λ_w must be real. The conclusion is that $(\lambda_v - \lambda_w)\langle \vec{v}, \vec{w}\rangle = 0$. The only way this could be true is if $\lambda_v = \lambda_w$ or if $\langle \vec{v}, \vec{w}\rangle = 0$. Since I assume the eigenvalues to be *distinct*, then the inner product of the two eigenvectors is 0. Specifically, if the matrix A has real components and is symmetric, then the dot product of two eigenvectors corresponding to two separate eigenvalues must vanish; i.e. they are orthogonal. In summary, I have proven the following theorem:

Theorem: If A is a symmetric matrix, then the eigenvalues are real, and eigenvectors corresponding to distinct eigenvalues are orthogonal.

Let's reexamine the matrix $A = \begin{pmatrix} 1 & 2 & 2 \\ 2 & 1 & 2 \\ 2 & 2 & 1 \end{pmatrix}$ in light of the discussion of this section. First, I found that 5 and -1 (with multiplicity 2) are the eigenvalues of this matrix. Indeed, they are both real as they should be since $A = A^T$. We also know that

$$\text{Ei}(\lambda = 5) = \text{Span}\left\{\begin{pmatrix} 1 \\ 1 \\ 1 \end{pmatrix}\right\} \text{ and } \text{Ei}(\lambda = -1) = \text{Span}\left\{\begin{pmatrix} -1 \\ 1 \\ 0 \end{pmatrix}, \begin{pmatrix} -1 \\ 0 \\ 1 \end{pmatrix}\right\}. \tag{5.6.7}$$

It is not hard to show the spanning vector of $\text{Ei}(\lambda = 5)$ is orthogonal to the spanning vectors of $\text{Ei}(\lambda = -1)$. In fact, the spaces are necessarily orthogonal in the following sense: $\text{Ei}(\lambda = 5)$ is a line through the origin in $\mathbb{R}^3$, and $\text{Ei}(\lambda = -1)$ is a plane through the origin. Where they intersect at the origin, the line and the plane form a right angle (see Fig. 5.5).

This matrix is diagonalizable. The eigenvectors defining the eigenspaces together form a basis

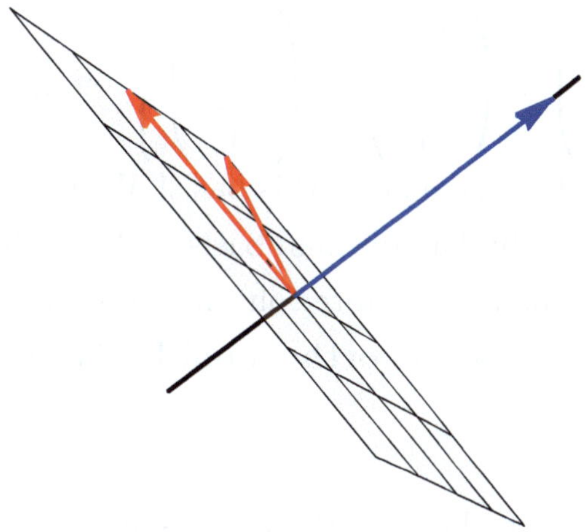

Figure 5.5: The plane corresponds to $\text{Ei}(\lambda = -1)$ (spanned by the red vectors) whereas the black line corresponds to $\text{Ei}(\lambda = 5)$ (spanned by the blue vector).

for $\mathbb{R}^3$, since they are linearly independent. As I'll show you in the next chapter, calculations using a basis in which *all* the basis vectors are orthogonal can be quite simple. We almost have that situation here. The problem is that the eigenvectors spanning $\text{Ei}(\lambda = -1)$ are not orthogonal as a direct calculation shows. It turns out that I can find a new spanning set for $\text{Ei}(\lambda = -1)$ and, hence, a new basis for $\mathbb{R}^3$ in which both $[A]_{\mathcal{B}'}$ is diagonal, and all of the basis vectors of $\mathcal{B}'$ are orthogonal. Since we don't want to change the fact the matrix A is diagonal in the current basis, our only freedom is in taking linear combinations of the vectors in the spanning set $\text{Ei}(\lambda = -1)$ to define a new basis. In fact, I can write

$$\text{Ei}(\lambda = -1) = \text{Span}\left\{\begin{pmatrix} -1 \\ 1 \\ 0 \end{pmatrix}, \begin{pmatrix} -1 \\ 0 \\ 1 \end{pmatrix}\right\} = \text{Span}\left\{\begin{pmatrix} -1 \\ 1 \\ 0 \end{pmatrix}, \begin{pmatrix} -1 \\ 0 \\ 1 \end{pmatrix} - \frac{1}{2}\begin{pmatrix} -1 \\ 1 \\ 0 \end{pmatrix}\right\} \tag{5.6.8}$$

or

$$\text{Ei}(\lambda = -1) = \text{Span}\left\{\begin{pmatrix} -1 \\ 1 \\ 0 \end{pmatrix}, \begin{pmatrix} -\frac{1}{2} \\ -\frac{1}{2} \\ 1 \end{pmatrix}\right\}. \tag{5.6.9}$$

The basis $\mathcal{B}' = \text{Span}\left\{ \begin{pmatrix} 1 \\ 1 \\ 1 \end{pmatrix}, \begin{pmatrix} -1 \\ 1 \\ 0 \end{pmatrix}, \begin{pmatrix} -\frac{1}{2} \\ -\frac{1}{2} \\ 1 \end{pmatrix} \right\}$ for $\mathbb{R}^3$ consists of mutually orthogonal eigenvectors of A, *and* A is diagonal in this basis! With a minor tweak to the spanning set for the $\lambda = -1$ eigenspace, I find an optimal basis with which to do calculations involving the symmetric matrix A. In the next chapter on inner product spaces, I will show you how to formalize this procedure to obtain an **orthogonal basis** for any symmetric matrix A via the Gram-Schmidt process.

Before we move on to the next chapter, there are still a few things to be learned from this symmetric matrix example. First, we know that we have exactly three real eigenvalues, counting multiplicity. Second, interestingly enough, the dimensions of the corresponding eigenspaces are precisely the multiplicities of the eigenvalues: $\text{Ei}(\lambda = -1)$ is two-dimensional, and $\text{Ei}(\lambda = 5)$ is one-dimensional, where $\lambda = -1$ has multiplicity 2 and $\lambda = 5$ has multiplicity 1. Third, it is possible to obtain a mutually orthogonal set of basis vectors diagonalizing A. It turns out that all of these observations are generally applicable to symmetric matrices.

The Spectral Theorem:[9] An $n \times n$ symmetric matrix A has the following characteristics:

1. A has exactly n real eigenvalues, counting multiplicities.
2. The dimension of the eigenspace corresponding to λ is equal to the multiplicity of λ.
3. The eigenspaces are mutually orthogonal, in the sense that eigenvectors corresponding to distinct eigenvalues are orthogonal.
4. There is a basis in which all of the basis vectors are mutually orthogonal, and A is represented by a diagonal matrix.

I have already proven properties 1 and 3 for you. I have yet to show you that property 4 is true, but let's just assume that is for the sake of argument. If I define $\mathcal{B}$ to be the standard basis and $\mathcal{B}'$ to be an orthogonal basis[10] then there is an invertible matrix relating the bases $C_{\mathcal{B}' \to \mathcal{B}}$ (that I will call V) such that $[A]_{\mathcal{B}} = A = VDV^{-1}$ where D is the diagonal

[9] The spectral theorem takes its name from *spectrum*, which is the term for the set of all eigenvalues of A.
[10] This just means that each pair of basis vectors is orthogonal.

matrix made up of the eigenvalues of A. There will be m copies of λ in D, if λ has multiplicity m. In the basis $\mathcal{B}'$, there are then m linearly independent vectors $\vec{x}_1, \ldots, \vec{x}_m$ for which $D\vec{x}_j = \lambda\vec{x}_j$ is true. Specifically, if $D_{jj} = \lambda$, define $\vec{x}_j$ by the coordinate vector $[\vec{x}_j]_{\mathcal{B}'} = \begin{pmatrix} 0 & \cdots & \underbrace{1}_{j\text{-th position}} & \cdots & 0 \end{pmatrix}^T$. Since the matrix V is invertible, it defines a linear, one-to-one and onto correspondence between $\text{Span}\{\vec{x}_1, \ldots, \vec{x}_m\}$ and the eigenspace $Ei(\lambda)$. Therefore, the two spaces are isomorphic and necessarily have the same dimension, namely the multiplicity of m.

Of course, this argument hinges on property 4 being true. I can demonstrate this property once I have told you about the Gram-Schmidt procedure. Before I get there, I need to tell you more about inner products and inner product spaces, so I turn to these topics now.

5.7 Chapter 5 Exercises

1. For this problem, assume that $\mathcal{B}$ consists of the standard basis vectors; i.e. $\mathcal{B} = \{\hat{e}_1, \hat{e}_2\}$. The vector sets $\mathcal{B}' = \left\{(2,-1)^T, (3,1)^T\right\}$, $\mathcal{B}'' = \left\{(-\frac{1}{2},0)^T, (\frac{1}{2},3)^T\right\}$, and $\mathcal{B}''' = \left\{(4,4)^T, (-3,-3)^T\right\}$ are all linearly independent and, therefore, form bases for the space $\mathbb{R}^2$. Compute $C_{\mathcal{B}\to\mathcal{B}'}$, $C_{\mathcal{B}\to\mathcal{B}''}$, and $C_{\mathcal{B}\to\mathcal{B}'''}$. Also compute $C_{\mathcal{B}'\to\mathcal{B}}$, $C_{\mathcal{B}''\to\mathcal{B}}$, and $C_{\mathcal{B}'''\to\mathcal{B}}$ directly and by computing the inverses of the matrices in the previous part of the problem. Given $A = \begin{pmatrix} 1 & -2 \\ 3 & 0 \end{pmatrix}$, compute $[A]_{\mathcal{B}'}$, $[A]_{\mathcal{B}''}$, and $[A]_{\mathcal{B}'''}$.

2. Given that $\lambda = 3$ is an eigenvalues for $A = \begin{pmatrix} 2 & -1 & -1 \\ -1 & 2 & -1 \\ -1 & -1 & 2 \end{pmatrix}$, find a basis for the corresponding eigenspace.

3. Diagonalize each of the following matrices if possible. If impossible, state why.

(a) $A = \begin{pmatrix} 2 & -1 & -1 \\ -1 & 2 & -1 \\ -1 & -1 & 2 \end{pmatrix}$; (b) $A = \begin{pmatrix} 5 & -6 & -6 \\ -1 & 4 & 2 \\ 3 & -6 & -4 \end{pmatrix}$; (c) $A = \begin{pmatrix} 6 & -3 & -2 \\ 4 & -1 & -2 \\ 10 & -5 & -3 \end{pmatrix}$;

(d) $A = \begin{pmatrix} -9 & 4 & 4 \\ -8 & 3 & 4 \\ -16 & 8 & 7 \end{pmatrix}$; (e) $A = \begin{pmatrix} 4 & 0 & -2 \\ 2 & 5 & 4 \\ 0 & 0 & 5 \end{pmatrix}$.

4. For what values of a and b (if any) is it possible to diagonalize $A = \begin{pmatrix} 0 & 0 & 0 & 0 \\ 1 & 0 & 0 & 0 \\ 0 & a & 0 & 0 \\ 0 & 0 & b & 0 \end{pmatrix}$?

5. For each of the following matrices, solve the first order linear system of differential equations $\frac{d\vec{x}}{dt} = A\vec{x}$. Describe (if possible) the behavior of the equilibrium $\vec{x} = \vec{0}$ near the origin (e.g., stable, unstable, spiral, etc.).

   (a) $A = \begin{pmatrix} 5 & -5 \\ 1 & 1 \end{pmatrix}$; (b) $A = \begin{pmatrix} 0 & 1 \\ -3 & 4 \end{pmatrix}$; $A = \begin{pmatrix} 2 & -6 \\ -1 & 2 \end{pmatrix}$.

6. Analyze the equilibrium $\vec{x} = (3, -1)^T$ of the first order *nonlinear* system $\frac{dx_1}{dt} = 2x_1 + 2x_1x_2$ and $\frac{dx_2}{dt} = -x_2x_1 + 3x_2$.

7. The matrix $A = \begin{pmatrix} 4 & 1 & -1 \\ 1 & 4 & -1 \\ -1 & -1 & 4 \end{pmatrix}$ is symmetric. Verify that the first three properties of the spectral theorem hold for this matrix. Can you find an orthogonal basis of eigenvectors?

8. Let A be a diagonalizable $n \times n$ matrix and $p(x)$ be the corresponding characteristic polynomial. Prove the identity $p(A) = 0$.

9. Let A be a diagonalizable $n \times n$ matrix with eigenvalues $\lambda_1, \ldots, \lambda_n$. Prove that $\det A = \lambda_1 \ldots \lambda_n$. Furthermore, define the **trace** of A (denoted $\mathrm{Tr}A$) to be the sum of the diagonal entries of A; i.e. $\mathrm{Tr}A = \sum_{i=1}^{n} a_{ii}$. The trace obeys the so-called cyclic property $\mathrm{Tr}\,(AB) = \mathrm{Tr}\,(BA)$. Show that for A diagonalizable, $\mathrm{Tr}A = \sum_{i=1}^{n} \lambda_i$.

10. Consider the nonlinear system $\frac{dx_1}{dt} = -x_1$, $\frac{dx_2}{dt} = x_2^2$. Without solving these equations, argue that general solutions with $x_2 \neq 0$ tend to move up and toward the x_2-axis. Show that the only equilibrium solution is $(0, 0)$. Linearize the system around this point. Solve the corresponding linear system, and compare the result to the observations above. What does this calculation tell you?

5.8 Chapter 5 References

If the reader wishes to explore the theory behind the theorems presented in this chapter or wishes to see more examples, I suggest consulting [5, 6]. The book [5] has a nice section devoted to applications of linear algebra to solving linear systems of differential equations. The book [6] develops the theory behind the spectral theorem in greater detail. Students interested in learning more about differential equations would do well to begin with [10].

Chapter 6

Inner Product Spaces

In a few words, inner product spaces are abstract vector spaces equipped with a machine called an **inner product** that takes in two vectors as input and returns a *complex* number as output. The two prominent examples we've seen so far are the dot product in $\mathbb{R}^n$ and the inner product $\langle \vec{u}, \vec{v} \rangle = \vec{v}^\dagger \vec{u}$ defined in the last chapter for $\vec{u}, \vec{v} \in \mathbb{C}^n$. To keep the notation simple, I'll use $\langle \boldsymbol{u}, \boldsymbol{v} \rangle$ for any inner product that we'll encounter in the rest of the book.

Definition: An **inner product space** consists of a complex vector space and a function (called an **inner product** $\langle \cdot, \cdot \rangle$) that takes, as input, two vectors $\boldsymbol{v}, \boldsymbol{w} \in V$ and returns a complex number $\langle \boldsymbol{v}, \boldsymbol{w} \rangle$. The inner product needs to satisfy the following properties:

1. $\langle c\boldsymbol{v}, \boldsymbol{w} \rangle = c\langle \boldsymbol{v}, \boldsymbol{w} \rangle \ , \ \langle \boldsymbol{v}, c\boldsymbol{w} \rangle = c^* \langle \boldsymbol{v}, \boldsymbol{w} \rangle$.
2. $\langle \boldsymbol{v}_1 + \boldsymbol{v}_2, \boldsymbol{w} \rangle = \langle \boldsymbol{v}_1, \boldsymbol{w} \rangle + \langle \boldsymbol{v}_2, \boldsymbol{w} \rangle \ , \ \langle \boldsymbol{v}, \boldsymbol{w}_1 + \boldsymbol{w}_2 \rangle = \langle \boldsymbol{v}, \boldsymbol{w}_1 \rangle + \langle \boldsymbol{v}, \boldsymbol{w}_2 \rangle$.
3. $\langle \boldsymbol{v}, \boldsymbol{w} \rangle^* = \langle \boldsymbol{w}, \boldsymbol{v} \rangle$.
4. $\langle \boldsymbol{v}, \boldsymbol{v} \rangle \geq 0 \ , \ \langle \boldsymbol{v}, \boldsymbol{v} \rangle = 0$ implies $\boldsymbol{v} = \boldsymbol{0}$.

By a complex vector space, I mean a set of elements together with operations of addition and scalar multiplication obeying all the rules of an ordinary vector spaces, but with the slight modification that I allow the scalars to be complex instead of requiring them to be real. The simplest example is $\mathbb{C}^n$ consisting of column vectors with n complex components. Such vector spaces arise often in physics. For instance, the complete description of a spin-1/2 particle utilizes a two-dimensional complex vector space. More down-to-earth applications arise in signal analysis (specifically, Fourier series), of which I'll have more to say later.

However, for the time being, I can say that a prominent example of a complex inner product space in the field of analysis is the set of all complex-valued continuous functions on the interval $[a, b]$ (denoted $\mathcal{C}([a, b])$) endowed with the inner product

$$\langle f, g\rangle = \int_a^b f(t)[g(t)]^* dt. \tag{6.0.1}$$

Of course, to call this an inner product, I would have to verify that all properties of an inner product hold. I will leave the details to the reader as it is a simple and illustrative example.

I used the dot product to define the length (norm) of a vector in $\mathbb{R}^n$. Since inner products generalize the dot product to abstract vector spaces, I can use the same basic concept to define the **norm** of a vector $\boldsymbol{v}$ in the inner product space V:

Definition: The **norm** of $\boldsymbol{v} \in V$ is $||\boldsymbol{v}|| = \sqrt{\langle \boldsymbol{v}, \boldsymbol{v}\rangle}$ for an inner product space V endowed with inner product $\langle\cdot,\cdot\rangle$.

For example, I can compute the length of $f(t) = \cos(t) \in \mathcal{C}([0, 2\pi])$ endowed with the inner product above:

$$||f|| = \sqrt{\int_0^{2\pi} f(t)[f(t)]^* dt} = \sqrt{\int_0^{2\pi} \cos^2(t) dt} = \sqrt{\pi}. \tag{6.0.2}$$

As a more interesting example, choose $g(t) = it \in \mathcal{C}([0, 1])$ where $i = \sqrt{-1}$:

$$||g|| = \sqrt{\int_0^1 g(t)[g(t)]^* dt} = \sqrt{\int_0^1 (it)\,(-it)\, dt} = \sqrt{\int_0^1 t^2 dt} = \frac{1}{\sqrt{3}}. \tag{6.0.3}$$

Given the way the inner product of a general inner product space V is defined, it is always the case that $\langle \boldsymbol{v}, \boldsymbol{v}\rangle$ is a nonnegative real number, and the norm is always well defined.

While I'm at it, I can also use the inner product to define the **distance** between two vectors in V. Recall in $\mathbb{R}^2$ that the norm of $\vec{v} - \vec{u}$ is given by $\sqrt{(v_1 - u_1)^2 + (v_2 - u_2)^2}$, which is precisely the distance between the two points (v_1, v_2) and (u_1, u_2) according to the Pythagorean theorem. So, in $\mathbb{R}^2$ (and in $\mathbb{R}^n$ as well) $||\vec{v} - \vec{u}||$ is the distance between the two vectors. This notion carries over immediately to the case of an inner product space V :

Definition: The **distance** between two vectors $\boldsymbol{v}, \boldsymbol{u}$ in an inner product space V is given by

$$\|\boldsymbol{v}-\boldsymbol{u}\| = \sqrt{\langle \boldsymbol{v}-\boldsymbol{u}, \boldsymbol{v}-\boldsymbol{u}\rangle}. \tag{6.0.4}$$

For example, the distance between $f(t) = i$ and $g(t) = \sqrt{t}$ in $\mathcal{C}([0,1])$ with the usual inner product is

$$\|f-g\| = \sqrt{\int_0^1 [i-\sqrt{t}][i-\sqrt{t}]^* dt} = \sqrt{\int_0^1 (1+t)dt} = \sqrt{\frac{3}{2}}. \tag{6.0.5}$$

Being able to define distances and lengths in an inner product space V is a great help in understanding the theorems I will prove and apply in this chapter.

The foci of this chapter on inner product spaces are the Best Approximation Theorem and the Gram-Schmidt Algorithm, both of which have numerous applications that I will touch on. In order to understand and apply these ideas, I first need to discuss some basic terminology. I begin with the concept of an **orthogonal set**.

6.1 Orthogonal Sets

Back when I introduced the dot product in $\mathbb{R}^n$, I mentioned the very useful formula $\vec{u}\cdot\vec{v} = \|\vec{u}\|\|\vec{v}\|\cos(\theta)$ where θ is the angle between the two vectors. For special pairs of vectors, the dot product will vanish, implying that the vectors make a right angle or are **orthogonal.** It is very useful to extend this idea of orthogonality to abstract inner product spaces:

Definition: Two vectors $\boldsymbol{u}, \boldsymbol{v} \in V$ in an inner product space are defined to be **orthogonal** if $\langle \boldsymbol{u}, \boldsymbol{v}\rangle = 0$.

For example, if I define $f(t) = \cos(t)$, $g(t) = \sin(t)$ to be vectors in the inner product space $\mathcal{C}([0,2\pi])$, then $\langle f, g\rangle = \int_0^{2\pi} \cos(t)\sin(t)dt = 0$ as a quick calculation shows. Therefore, sine and cosine are orthogonal in this space.

In general, orthogonal vectors are very nice to work with. For example, in many applications it is most convenient to work with a basis for our inner product space in which all of the basis vectors are pair-wise orthogonal. The standard basis in $\mathbb{R}^n$ is an example.

Definition: A set of vectors $\{\boldsymbol{v}_1, \ldots, \boldsymbol{v}_p\}$ is called an **orthogonal set** if every pair of vectors from this set is an orthogonal pair; i.e. $\langle \boldsymbol{v}_i, \boldsymbol{v}_j \rangle = 0$ for all $i \neq j$. If, in addition, the vectors also have unit length, then the vectors are called **orthonormal**.

Generally, if we have an orthogonal set, we can always construct an orthonormal set just by **normalizing** all of the vectors in the set. Normalization just means that we replace a vector $\boldsymbol{u}$ with $\frac{1}{\|\boldsymbol{u}\|}\boldsymbol{u}$, which has unit length by construction.

To give you a sense of why an **orthogonal basis** (i.e. a basis that is also an orthogonal set) is nice to work with, I'll revisit an old problem: given that $\boldsymbol{x} \in V$ and $\{\boldsymbol{v}_1 , \ldots , \boldsymbol{v}_p\}$ is an orthogonal basis for V, find the coordinates of $\boldsymbol{x}$ with respect to this basis. When our space V is $\mathbb{R}^n$, we can set up an augmented matrix consisting of the columns representing $\{\boldsymbol{v}_1 , \ldots , \boldsymbol{v}_p\}$, plus an additional column for $\boldsymbol{x}$. Row reducing this matrix using Gauss-Jordan, we can find the coordinates after some tedious calcuation. If our basis is orthogonal, there is a faster method at our disposal, often referred to as Fourier's trick:

Write $\boldsymbol{x} = c_1\boldsymbol{v}_1 + \cdots + c_p\boldsymbol{v}_p$ where the weights are the coordinates of $\boldsymbol{x}$, which are currently unknown. I exploit the fact that $\langle \boldsymbol{v}_i, \boldsymbol{v}_j \rangle = 0$ to isolate each of the weights in turn. For instance, suppose I want to calculate c_1. If I take the inner product of $\boldsymbol{x}$ with $\boldsymbol{v}_1$, I find

$$\langle \boldsymbol{x}, \boldsymbol{v}_1 \rangle = \langle c_1\boldsymbol{v}_1 + \cdots + c_p\boldsymbol{v}_p, \boldsymbol{v}_1 \rangle = c_1\langle \boldsymbol{v}_1, \boldsymbol{v}_1 \rangle, \tag{6.1.1}$$

since all of the other inner products vanish! Therefore, $c_1 = \frac{\langle \boldsymbol{x}, \boldsymbol{v}_1 \rangle}{\langle \boldsymbol{v}_1, \boldsymbol{v}_1 \rangle}$. In fact, I have just demonstrated a very useful theorem:

Theorem: Suppose $\{\boldsymbol{v}_1 , \ldots , \boldsymbol{v}_p\}$ is an orthogonal basis for the inner product space V. If $\boldsymbol{x} \in V$, then $\boldsymbol{x} = c_1\boldsymbol{v}_1 + \cdots + c_p\boldsymbol{v}_p$ where

$$c_j = \frac{\langle \boldsymbol{x}, \boldsymbol{v}_j \rangle}{\langle \boldsymbol{v}_j, \boldsymbol{v}_j \rangle} \tag{6.1.2}$$

are the unique weights for $\boldsymbol{x}$.

Example: Recall that $\left\{\vec{v}_1 = \begin{pmatrix} 1 & 1 & 1 \end{pmatrix}^T, \vec{v}_2 = \begin{pmatrix} -1 & 1 & 0 \end{pmatrix}^T, \vec{v}_3 = \begin{pmatrix} -\frac{1}{2} & -\frac{1}{2} & 1 \end{pmatrix}^T\right\}$ form an orthogonal basis for $\mathbb{R}^3$ in which $A = \begin{pmatrix} 1 & 2 & 2 \\ 2 & 1 & 2 \\ 2 & 2 & 1 \end{pmatrix}$ is diagonal. In particular, it must be possible to write $\hat{e}_3$ as a unique linear combination of $\vec{v}_1, \vec{v}_2$ and $\vec{v}_3$. According to the theorem above, the weights are given by $c_j = \frac{\langle \hat{e}_3, \vec{v}_j \rangle}{\langle \vec{v}_j, \vec{v}_j \rangle}$ where the inner product is just the standard dot product in $\mathbb{R}^3$. It easy to show that $\langle \hat{e}_3, \vec{v}_1 \rangle = 1$, $\langle \hat{e}_3, \vec{v}_2 \rangle = 0$, and $\langle \hat{e}_3, \vec{v}_3 \rangle = 1$. Furthermore, $\langle \vec{v}_1, \vec{v}_1 \rangle = 3$, $\langle \vec{v}_2, \vec{v}_2 \rangle = 2$, and $\langle \vec{v}_3, \vec{v}_3 \rangle = \frac{3}{2}$. The claim, then, is that $c_1 = \frac{1}{3}$, $c_2 = 0$, and $c_3 = \frac{2}{3}$. It is a simple matter to show that $\hat{e}_3 = \frac{1}{3}\vec{v}_1 + \frac{2}{3}\vec{v}_3$ indeed holds true.

Example: The set of functions $\{\sin(t), \cos(t), \sin(2t), \cos(2t)\}$ form an orthogonal basis for the subspace $W = \text{Span}\{\sin(t), \cos(t), \sin(2t), \cos(2t)\}$ of $C([0, 2\pi])$. I claim that the function $\sin(t)\cos(t)$ belongs to W as well.[1] If I write $\sin(t)\cos(t) = c_1 \sin(t) + c_2 \cos(t) + c_3 \sin(2t) + c_4 \cos(2t)$, the weights are given by:

$$c_1 = \frac{\langle \sin(t)\cos(t), \sin(t) \rangle}{\langle \sin(t), \sin(t) \rangle} = \frac{\int_0^{2\pi} \sin^2(t)\cos(t)dt}{\int_0^{2\pi} \sin^2(t)dt} = 0. \tag{6.1.3}$$

$$c_2 = \frac{\langle \sin(t)\cos(t), \cos(t) \rangle}{\langle \cos(t), \cos(t) \rangle} = \frac{\int_0^{2\pi} \sin(t)\cos^2(t)dt}{\int_0^{2\pi} \cos^2(t)dt} = 0. \tag{6.1.4}$$

$$c_3 = \frac{\langle \sin(t)\cos(t), \sin(2t) \rangle}{\langle \sin(2t), \sin(2t) \rangle} = \frac{\int_0^{2\pi} \sin(t)\cos(t)\sin(2t)dt}{\int_0^{2\pi} \sin^2(2t)dt} = \frac{1}{2}. \tag{6.1.5}$$

$$c_4 = \frac{\langle \sin(t)\cos(t), \cos(2t) \rangle}{\langle \cos(2t), \cos(2t) \rangle} = \frac{\int_0^{2\pi} \sin(t)\cos(t)\cos(2t)dt}{\int_0^{2\pi} \cos^2(2t)dt} = 0. \tag{6.1.6}$$

Therefore, $\sin(t)\cos(t) = \frac{1}{2}\sin(2t)$.

[1]This follows from the trigonometric identity $2\sin(t)\cos(t) = \sin(2t)$, but let's pretend like we don't know this fact.

6.2 Orthogonal Projection

The dot product in $\mathbb{R}^n$ allows us to define a notion of **orthogonal projection** that is incredibly useful not only in the applications I'll cover here, but also in physics and engineering where it is important to understand forces and stresses along specific directions of a system. This notion can, of course, be generalized to inner product spaces V, but I'll first present the general idea in $\mathbb{R}^2$ where it is easy to visualize what's going on. For the following discussion,

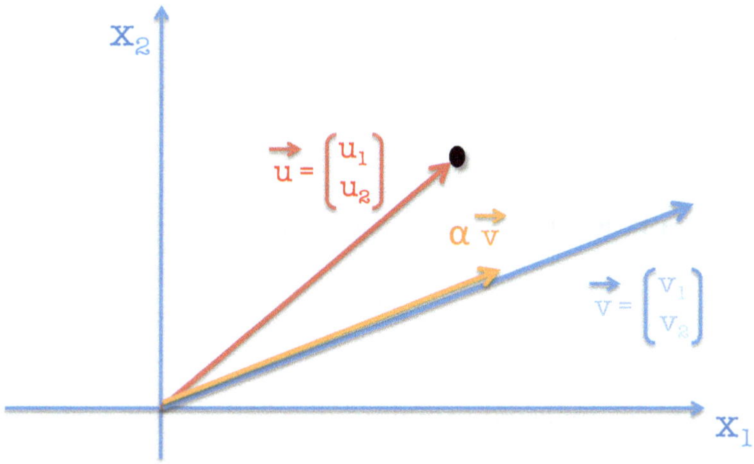

Figure 6.1: A vector $\vec{v}$ (blue) in the plane $\mathbb{R}^2$ may be stretched, reflected, or shrunk at will by multiplying by α (orange). A properly chosen α can make the distance between the tip of the orange vector $\alpha\vec{v}$ and the red vector $\vec{u}$ as small possible.

refer to Fig. 6.1. Suppose that we have two nonzero vectors in $\mathbb{R}^2$. Call these vectors $\vec{u}$ and $\vec{v}$, and suppose, for simplicity, that the vectors are linearly independent. By multiplying $\vec{v}$ by some constant α, I can stretch, shrink, and reflect $\vec{v}$ as I please. For reasons that will be made clear in a moment, suppose I want to choose α so that $\alpha\vec{v}$ is like the shadow of $\vec{u}$ cast upon $\vec{v}$ by a distant light source in the plane. More precisely, suppose I want to find α (if it exists) so that the end of the arrow representing $\alpha\vec{v}$ is as close as possible to the end of the arrow representing $\vec{u}$. Remember, $\vec{u}$ and $\vec{v}$ (and hence its direction) are fixed quantities; I only have control over α.

The distance will be given by the expression $||\vec{u} - \alpha\vec{v}||$ for any choice of α. So, if I plug in a certain value for α, out comes the distance between $\alpha\vec{v}$ and $\vec{u}$ assuming I know the components of the vectors. In other words, I have a function of a real variable (α) with a real-valued output (the distance between $\alpha\vec{v}$ and $\vec{u}$). Call this function $f(\alpha)$. As α approaches $\pm\infty$, $f(\alpha)$ approaches positive infinity. For some finite value of α, f is a small as possible, since f is necessarily bounded below (e.g., it is always nonnegative). To find the minimum, I can take the derivative of $f(\alpha)$ with respect to α and set it equal to 0:

$$f'(\alpha) = \frac{(\alpha v_1 - u_1)v_1 + (\alpha v_2 - u_2)v_2}{\sqrt{(u_1 - \alpha v_1)^2 + (u_2 - \alpha v_2)^2}} = 0. \tag{6.2.1}$$

This equation has the unique answer

$$\alpha = \frac{u_1 v_1 + u_2 v_2}{v_1^2 + v_2^2} = \frac{\vec{u}\cdot\vec{v}}{\vec{v}\cdot\vec{v}}. \tag{6.2.2}$$

I define the **orthogonal projection** of $\vec{u}$ onto $\vec{v}$ to be

$$\operatorname{proj}_{\vec{v}} \vec{u} = \left(\frac{\vec{u}\cdot\vec{v}}{\vec{v}\cdot\vec{v}}\right)\vec{v}. \tag{6.2.3}$$

The term "projection" should make you think of $\vec{u}$'s shadow being cast on $\vec{v}$ from a distant light source. Alternatively (and more precisely), the projection of $\vec{u}$ onto $\vec{v}$ is a measure of how much $\vec{u}$ runs along $\vec{v}$; i.e. the component of $\vec{u}$ that runs along the $\vec{v}$ direction. In point of fact, I can rearrange the equation defining the orthogonal projection so that

$$\operatorname{proj}_{\vec{v}} \vec{u} = \left(\frac{\vec{u}\cdot\vec{v}}{\vec{v}\cdot\vec{v}}\right)\vec{v} = \left(\vec{u}\cdot\frac{\vec{v}}{\sqrt{\vec{v}\cdot\vec{v}}}\right)\frac{\vec{v}}{\sqrt{\vec{v}\cdot\vec{v}}} = (\vec{u}\cdot\hat{v})\,\hat{v}, \tag{6.2.4}$$

where $\hat{v}$ is the unit vector pointing in the direction of $\vec{v}$ obtained after normalizing $\vec{v}$. If I write $\vec{u} = u_1\hat{e}_1 + u_2\hat{e}_2$, then it is clear that u_j measures how much of $\vec{u}$ runs along the $\hat{e}_j$-direction. Moreover, $\operatorname{proj}_{\hat{e}_j} \vec{u} = (\vec{u}\cdot\hat{e}_j)\,\hat{e}_j = u_j\hat{e}_j$, so we see that the projection of $\vec{u}$ onto the standard basis vector $\hat{e}_j$ gives us the component of $\vec{u}$ in the $\hat{e}_j$-direction. By analogy, $\operatorname{proj}_{\vec{v}} \vec{u} = (\vec{u}\cdot\hat{v})\,\hat{v}$ gives us the component of $\vec{u}$ in the $\vec{v}$-direction.

You may wonder where the "orthogonal" in orthogonal projection comes from. If I define $\vec{z} = \vec{u} - \operatorname{proj}_{\vec{v}} \vec{u}$ (see Fig. 6.2), then it is a simple calculation to show that

$$\vec{v}\cdot\vec{z} = \vec{v}\cdot\vec{u} - \left(\frac{\vec{u}\cdot\vec{v}}{\vec{v}\cdot\vec{v}}\right)(\vec{v}\cdot\vec{v}) = 0, \tag{6.2.5}$$

so $\vec{z}$ is necessarily orthogonal to $\vec{v}$. The formula for orthogonal projection in $\mathbb{R}^2$ doesn't

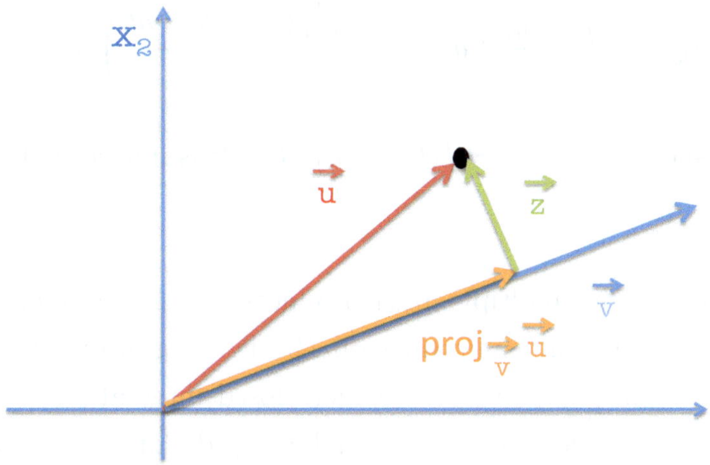

Figure 6.2: A vector $\vec{u}$ (red) in the plane $\mathbb{R}^2$ is projected onto the vector $\vec{v}$ (blue). The projection vector $\text{proj}_{\vec{v}}\ \vec{u}$ (orange) is the component of $\vec{u}$ in the $\vec{v}$-direction. The vector $\vec{z} = \vec{u} - \text{proj}_{\vec{v}}\ \vec{u}$ (green) is orthogonal to $\vec{v}$; i.e. they form a right angle.

depend on the number of components at all, just on the definition of the dot product. So it is easy to generalize the definition to vectors in $\mathbb{R}^n$. In fact, if we replace dot product with inner product, the definition carries over immediately to general inner product spaces:

Definition: Given $\boldsymbol{u}, \boldsymbol{v} \in V$ where V is an inner product space, the **orthogonal projection** of $\boldsymbol{u}$ onto $\boldsymbol{v} \neq \boldsymbol{0}$ is

$$\text{proj}_{\boldsymbol{v}}\ \boldsymbol{u} = \left(\frac{\langle \boldsymbol{u}, \boldsymbol{v}\rangle}{\langle \boldsymbol{v}, \boldsymbol{v}\rangle}\right) \boldsymbol{v}. \tag{6.2.6}$$

Note that $\boldsymbol{z} = \boldsymbol{u} - \text{proj}_{\boldsymbol{v}}\ \boldsymbol{u}$ satisfies

$$\langle \boldsymbol{z}, \boldsymbol{v}\rangle = \langle \boldsymbol{u} - \text{proj}_{\boldsymbol{v}}\ \boldsymbol{u}, \boldsymbol{v}\rangle = \langle \boldsymbol{u}, \boldsymbol{v}\rangle - \left(\frac{\langle \boldsymbol{u}, \boldsymbol{v}\rangle}{\langle \boldsymbol{v}, \boldsymbol{v}\rangle}\right) \langle \boldsymbol{v}, \boldsymbol{v}\rangle = 0 \tag{6.2.7}$$

by the basic properties of the inner product $\langle \cdot, \cdot \rangle$.

Example: One way of defining an inner product on P_n is by use of the integral: $\langle p(t), q(t)\rangle = \int_{-1}^{1} p(t)q(t)dt$. Here, I have singled out the interval $[-1, 1]$ just for computational convenience and for no other reason. Using this inner product, it is possible to compute the orthogonal projection of $f(t) = t^2 + 2t$ onto $g(t) = t$ where $f, g \in P_2$;

$$\text{proj}_g\ f = \left(\frac{\langle f, g\rangle}{\langle g, g\rangle}\right) g(t) = \left(\frac{\int_{-1}^{1} (t^2 + 2t)\, t\ dt}{\int_{-1}^{1} t^2 dt}\right) t = 2t. \tag{6.2.8}$$

A simple calculation shows that $z(t) = f(t) - \text{proj}_g\ f = t^2$ is indeed orthogonal to $g(t)$ under this inner product.

It is certainly possible to define multiple inner products on a given inner product space.[2] For example, I can produce a new inner product on P_n via integration simply by changing the interval of integration from $[-1, 1]$ to some other closed interval $[a, b]$. A more interesting inner product on P_n can be made through the following definition: choose $t_0, t_1, \ldots, t_n \in \mathbb{R}$ to be fixed points and define

$$(f, g) = f(t_0)g(t_0) + \cdots + f(t_n)g(t_n). \tag{6.2.9}$$

Here, I use the notation (f, g) so as not to confuse this inner product with $\langle f, g\rangle = \int_{-1}^{1} f(t)g(t)dt$. All of the properties of an inner product are easily seen to be satisfied, save one: some work is needed to show that $(f, f) \geq 0$ and $(f, f) = 0$ if and only if $f(t) = 0$. By definition, $(f, f) = f(t_0)^2 + \cdots + f(t_n)^2$. Since this is a sum of squares of real numbers, $(f, f) \geq 0$ always. If $(f, f) = 0$, however, it must be true that all terms in the sum $f(t_0)^2 + \cdots + f(t_n)^2$ vanish: if not, then the sum necessarily has to be strictly bigger than 0. Therefore, $(f, f) = 0$ implies that $f(t_0) = f(t_1) = \cdots = f(t_n) = 0$. From this, we see that $f(t)$ is a polynomial of degree no larger than n, yet has $n + 1$ roots. The only way this could be true is if $f(t) = 0$ identically. This statement follows from the fundamental theorem of algebra.

For our example, I'll use $t_0 = -1, t_1 = 0, t_2 = 2$. Using this inner product, $\text{proj}_g\ f = \left(\frac{(f,g)}{(g,g)}\right) g(t)$, so

$$\text{proj}_g\ f = \left(\frac{((-1)^2 + 2(-1))(-1) + (0^2 + 2 \cdot 0)0 + (2^2 + 2 \cdot 2)(1)}{(-1)^2 + 0^2 + 2^2}\right) t = \frac{9}{5}t, \tag{6.2.10}$$

so the orthogonal projection is *not* the same as in the previous calculation. This example demonstrates that length, orthogonality, orthogonal projection, etc. are all tied to the spe-

[2] In fact, if $\langle \cdot, \cdot \rangle$ is an inner product on V, then $c\langle \cdot, \cdot \rangle$ is also an inner product, as long as $c > 0$ and $c \in \mathbb{R}$.

cific choice of inner product you use: change the inner product, and you will, in general, change the basic geometry of your inner product space.

Example: Recall that the symmetric matrix $A = \begin{pmatrix} 1 & 2 & 2 \\ 2 & 1 & 2 \\ 2 & 2 & 1 \end{pmatrix}$ has eigenspaces

$$\text{Ei}(\lambda = 5) = \text{Span}\left\{\begin{pmatrix} 1 & 1 & 1 \end{pmatrix}^T\right\} \text{ and } \text{Ei}(\lambda = -1) = \text{Span}\left\{\begin{pmatrix} -1 & 1 & 0 \end{pmatrix}^T, \begin{pmatrix} -1 & 0 & 1 \end{pmatrix}^T\right\}. \tag{6.2.11}$$

Previously, I argued that a new linear combination of the vectors in the spanning set of $\text{Ei}(\lambda = -1)$ gives rise to an orthogonal basis of eigenvectors for eigenvalue $\lambda = -1$. I am now in a position to show you how this arrangement can be computed. Keep $\begin{pmatrix} -1 & 1 & 0 \end{pmatrix}^T$, and modify $\begin{pmatrix} -1 & 0 & 1 \end{pmatrix}^T$ to obtain an orthogonal basis.[3] By definition of a span, it is true that

$$\text{Ei}(\lambda = -1) = \text{Span}\left\{\begin{pmatrix} -1 & 1 & 0 \end{pmatrix}^T, \begin{pmatrix} -1 & 0 & 1 \end{pmatrix}^T + \alpha\begin{pmatrix} -1 & 1 & 0 \end{pmatrix}^T\right\} \tag{6.2.12}$$

for any choice of $\alpha \in \mathbb{R}$. If I choose

$$\alpha = -\left(\frac{\begin{pmatrix} -1 & 0 & 1 \end{pmatrix}^T \cdot \begin{pmatrix} -1 & 1 & 0 \end{pmatrix}^T}{\begin{pmatrix} -1 & 1 & 0 \end{pmatrix}^T \cdot \begin{pmatrix} -1 & 1 & 0 \end{pmatrix}^T}\right) = -\frac{1}{2} \tag{6.2.13}$$

so that the second spanning vector is orthogonal to the first, then

$$\text{Ei}(\lambda = -1) = \text{Span}\left\{\begin{pmatrix} -1 & 1 & 0 \end{pmatrix}^T, \begin{pmatrix} -\frac{1}{2} & -\frac{1}{2} & 1 \end{pmatrix}^T\right\}, \tag{6.2.14}$$

which is precisely the orthogonal basis I found earlier! This simple example shows how orthogonal projection can be used to convert a basis into an orthogonal basis. The general procedure for taking any arbitrary basis in an inner product space and constructing a new orthogonal basis is referred to as the **Gram-Schmidt process**, to which I turn our attention after a brief interlude.

[3]There is nothing special about this choice: I could reverse the roles of the two vectors and obtain a perfectly valid, though possibly different, orthogonal basis for the eigenspace.

The Cauchy-Schwarz Inequality

One of the more important inequalities of mathematics is the **Cauchy-Schwarz Inequality**:

Cauchy-Schwarz Inequality: Given $\boldsymbol{u}, \boldsymbol{v} \in V$, $|\langle \boldsymbol{u}, \boldsymbol{v} \rangle| \leq \|\boldsymbol{u}\| \|\boldsymbol{v}\|$.

We know from past experience that $\vec{u} \cdot \vec{v} = \|\vec{u}\| \|\vec{v}\| \cos(\theta)$ for vectors in $\mathbb{R}^n$. Since cosine is bounded between -1 and 1, the inequality follows trivially. The Cauchy-Schwarz Inequality can be considered a generalization of this formula for a general inner product space V. To prove the inequality, I have to be a bit more clever in my approach. Given $\boldsymbol{u}$ and $\boldsymbol{v}$, allow me to define $c = \langle \boldsymbol{u}, \boldsymbol{v} \rangle \in \mathbb{C}$, and consider the inner product $\langle \boldsymbol{u} - c\boldsymbol{v}, \boldsymbol{u} - c\boldsymbol{v} \rangle$, which is necessarily a nonnegative real number by the properties of $\langle \cdot, \cdot \rangle$. Without any real loss, I'll assume for the moment that both vectors are unit vectors. Then

$$\langle \boldsymbol{u} - c\boldsymbol{v}, \boldsymbol{u} - c\boldsymbol{v} \rangle = \langle \boldsymbol{u}, \boldsymbol{u} \rangle + |c|^2 \langle \boldsymbol{v}, \boldsymbol{v} \rangle - c^* \langle \boldsymbol{u}, \boldsymbol{v} \rangle - c \langle \boldsymbol{v}, \boldsymbol{u} \rangle = 1 - |\langle \boldsymbol{u}, \boldsymbol{v} \rangle|^2 \geq 0. \quad (6.2.15)$$

It follows that $|\langle \boldsymbol{u}, \boldsymbol{v} \rangle| \leq 1 = \|\boldsymbol{u}\| \|\boldsymbol{v}\|$ when the vectors are unit vectors. If they are not unit vectors, then $\boldsymbol{u}/\|\boldsymbol{u}\|$ and $\boldsymbol{v}/\|\boldsymbol{v}\|$ can be used instead in which case the above inequality becomes $\langle \boldsymbol{u}/\|\boldsymbol{u}\|, \boldsymbol{v}/\|\boldsymbol{v}\| \rangle \leq 1$, from which the Cauchy-Schwarz inequality follows immediately.

A consequence of this inequality is the **Triangle Inequality:** $\|\boldsymbol{u} + \boldsymbol{v}\| \leq \|\boldsymbol{u}\| + \|\boldsymbol{v}\|$. If you imagine the vectors as being represented by arrows, then the triple $\boldsymbol{u}, \boldsymbol{v}, \boldsymbol{u} + \boldsymbol{v}$ defines a triangle, hence the name. The triangle inequality is very useful in proving theorems in multivariable calculus and advanced math. It is amazing that such a useful tool can be proven using just basic linear algebra! The proof goes something like this:

$$\|\boldsymbol{u} + \boldsymbol{v}\|^2 = \|\boldsymbol{u}\|^2 + \|\boldsymbol{v}\|^2 + \langle \boldsymbol{u}, \boldsymbol{v} \rangle + \langle \boldsymbol{v}, \boldsymbol{u} \rangle = \|\boldsymbol{u}\|^2 + \|\boldsymbol{v}\|^2 + 2\mathrm{Re}\langle \boldsymbol{u}, \boldsymbol{v} \rangle. \quad (6.2.16)$$

Since the real part of a complex number is always bounded above by the absolute value of that number, we have $\mathrm{Re}\langle \boldsymbol{u}, \boldsymbol{v} \rangle \leq |\langle \boldsymbol{u}, \boldsymbol{v} \rangle| \leq \|\boldsymbol{u}\| \|\boldsymbol{v}\|$ by the Cauchy-Schwarz inequality. Therefore,

$$\|\boldsymbol{u} + \boldsymbol{v}\|^2 \leq \|\boldsymbol{u}\|^2 + \|\boldsymbol{v}\|^2 + 2\|\boldsymbol{u}\| \|\boldsymbol{v}\| = (\|\boldsymbol{u}\| + \|\boldsymbol{v}\|)^2, \quad (6.2.17)$$

proving the inequality.

6.3 The Gram-Schmidt Process

We know that orthogonal bases are highly desirable, since they are so nice to work with. Ordinarily, though, we are able to construct a basis for an inner product space that is not necessarily orthogonal. For instance, we have computed bases for both Col A and Nul A for an $m \times n$ matrix A using the Gauss-Jordan algorithm in the past. However, nothing guarantees that these bases are orthogonal.

Based on what we saw in the last section, it makes sense to appeal to the orthogonal projection procedure to construct an orthogonal basis out of a given basis. Before I state the general theorem, I'll give you an overview of how the process works. First, you need a basis (constructed by any means) to initiate the algorithm. Call this basis $\{\boldsymbol{v}_1, \ldots, \boldsymbol{v}_p\}$. We are free to take appropriate linear combinations of the vectors of this basis to obtain new bases. Just like in the example ending the last section, I can replace $\boldsymbol{v}_2$ with $\boldsymbol{w}_2 = \boldsymbol{v}_2 - \text{proj}_{\boldsymbol{v}_1}\boldsymbol{v}_2$, which is necessarily orthogonal to $\boldsymbol{v}_1$. For consistency, I'll also define $\boldsymbol{w}_1 = \boldsymbol{v}_1$, since I want to label the new basis with $\boldsymbol{w}$'s. Next, I can take $\boldsymbol{v}_3$ and orthogonally project onto both $\boldsymbol{w}_1$ and $\boldsymbol{w}_2$; the vector $\boldsymbol{w}_3 = \boldsymbol{v}_3 - \text{proj}_{\boldsymbol{w}_1}\boldsymbol{v}_3 - \text{proj}_{\boldsymbol{w}_2}\boldsymbol{v}_3$ is necessarily orthogonal to both $\boldsymbol{w}_1$ and $\boldsymbol{w}_2$ since $\boldsymbol{w}_1$ and $\boldsymbol{w}_2$ are orthogonal by construction. I continue this process of orthogonally projecting vector $\boldsymbol{v}_j$ onto the previously constructed $\{\boldsymbol{w}_1, \ldots \boldsymbol{w}_j\}$ until an orthogonal basis is constructed.

Theorem (The Gram-Schmidt Algorithm): Given a basis $\{\boldsymbol{v}_1, \ldots, \boldsymbol{v}_p\}$ for an inner product space V, define

1. $\boldsymbol{w}_1 = \boldsymbol{v}_1$
2. $\boldsymbol{w}_2 = \boldsymbol{v}_2 - \left(\frac{\langle \boldsymbol{v}_2, \boldsymbol{w}_1\rangle}{\langle \boldsymbol{w}_1, \boldsymbol{w}_1\rangle}\right)\boldsymbol{w}_1$
3. $\boldsymbol{w}_3 = \boldsymbol{v}_3 - \left(\frac{\langle \boldsymbol{v}_3, \boldsymbol{w}_1\rangle}{\langle \boldsymbol{w}_1, \boldsymbol{w}_1\rangle}\right)\boldsymbol{w}_1 - \left(\frac{\langle \boldsymbol{v}_3, \boldsymbol{w}_2\rangle}{\langle \boldsymbol{w}_2, \boldsymbol{w}_2\rangle}\right)\boldsymbol{w}_2$
4. $\vdots$
5. $\boldsymbol{w}_p = \boldsymbol{v}_p - \left(\frac{\langle \boldsymbol{v}_p, \boldsymbol{w}_1\rangle}{\langle \boldsymbol{w}_1, \boldsymbol{w}_1\rangle}\right)\boldsymbol{w}_1 - \left(\frac{\langle \boldsymbol{v}_p, \boldsymbol{w}_2\rangle}{\langle \boldsymbol{w}_2, \boldsymbol{w}_2\rangle}\right)\boldsymbol{w}_2 - \cdots - \left(\frac{\langle \boldsymbol{v}_p, \boldsymbol{w}_{p-1}\rangle}{\langle \boldsymbol{w}_{p-1}, \boldsymbol{w}_{p-1}\rangle}\right)\boldsymbol{w}_{p-1}.$

Then, $\{\boldsymbol{w}_1, \ldots, \boldsymbol{w}_p\}$ is an orthogonal basis for V.

As I've defined the vectors $\boldsymbol{w}_j$, it is straightforward to show that $\boldsymbol{w}_j$ is orthogonal to $\boldsymbol{w}_1, \ldots, \boldsymbol{w}_{j-1}$ for $j > 1$. Therefore, the set $\{\boldsymbol{w}_1, \ldots, \boldsymbol{w}_p\}$ is orthogonal. Furthermore, if *any* set of vectors $\{\boldsymbol{w}_1, \ldots, \boldsymbol{w}_p\}$ is orthogonal, then the set is linearly independent automatically; if

$$\mathbf{0} = c_1 \boldsymbol{w}_1 + \cdots + c_p \boldsymbol{w}_p, \tag{6.3.1}$$

then Fourier's trick tells us that $\langle \mathbf{0}, \boldsymbol{w}_j \rangle = 0 = c_j \langle \boldsymbol{w}_j, \boldsymbol{w}_j \rangle$, implying $c_j = 0$. Therefore, an orthogonal set is automatically linearly independent. Since there are exactly p vectors in the set $\{\boldsymbol{w}_1, \ldots, \boldsymbol{w}_p\}$ constructed via the Gram-Schmidt process above, and the dimension of V is p, the set $\{\boldsymbol{w}_1, \ldots, \boldsymbol{w}_p\}$ is an orthogonal basis for V.

Example: The vector space

$$W = \text{Span}\left\{ \vec{v}_1 = \begin{pmatrix} -1 & -1 & 0 & 0 \end{pmatrix}^T, \vec{v}_2 = \begin{pmatrix} 1 & -1 & 1 & 1 \end{pmatrix}^T, \vec{v}_3 = \begin{pmatrix} 0 & 0 & 1 & -2 \end{pmatrix}^T \right\} \tag{6.3.2}$$

inherits an inner product from $\mathbb{R}^4$, since W is a subspace of this space. Specifically, W is an inner product space with the usual dot product. The vectors in the spanning set are linearly independent and, therefore, serve as a basis for W. I'll use the Gram-Schmidt algorithm to find an *orthonormal* basis for W. Start by finding an orthogonal basis:

$$\vec{w}_1 = \vec{v}_1 = \begin{pmatrix} -1 \\ -1 \\ 0 \\ 0 \end{pmatrix}, \quad \vec{w}_2 = \vec{v}_2 - \left(\frac{\langle \vec{v}_2, \vec{w}_1 \rangle}{\langle \vec{w}_1, \vec{w}_1 \rangle} \right) \vec{w}_1 = \begin{pmatrix} 1 \\ -1 \\ 1 \\ 1 \end{pmatrix} \tag{6.3.3}$$

and

$$\vec{w}_3 = \vec{v}_3 - \left(\frac{\langle \vec{v}_3, \vec{w}_1 \rangle}{\langle \vec{w}_1, \vec{w}_1 \rangle} \right) \vec{w}_1 - \left(\frac{\langle \vec{v}_3, \vec{w}_2 \rangle}{\langle \vec{w}_2, \vec{w}_2 \rangle} \right) \vec{w}_2 = \vec{v}_3 - 0\vec{w}_1 - \left(-\frac{1}{4} \right) \vec{w}_2 = \begin{pmatrix} \frac{1}{4} \\ -\frac{1}{4} \\ \frac{5}{4} \\ -\frac{7}{4} \end{pmatrix}. \tag{6.3.4}$$

A direct calculation shows that all of the vectors are indeed orthogonal. To complete the problem, I simply have to normalize each vector to find an orthonormal basis:

$$W = \text{Span}\left\{\begin{pmatrix} -\frac{1}{\sqrt{2}} & -\frac{1}{\sqrt{2}} & 0 & 0 \end{pmatrix}^T, \begin{pmatrix} \frac{1}{2} & -\frac{1}{2} & \frac{1}{2} & \frac{1}{2} \end{pmatrix}^T, \begin{pmatrix} \frac{1}{\sqrt{76}} & -\frac{1}{\sqrt{76}} & \frac{5}{\sqrt{76}} & -\frac{7}{\sqrt{76}} \end{pmatrix}^T\right\}. \tag{6.3.5}$$

6.4 Orthogonal Decomposition

One immediate application of the Gram-Schmidt algorithm is proving the following fact: any finite-dimensional inner product space V has an orthogonal basis. The proof is simple: since the space is finite-dimensional, there is a basis for V that may be used in the Gram-Schmidt process to obtain an orthogonal basis. In turn, a consequence of this fact is the following:

Theorem: (Orthogonal Decomposition) Let V be an inner product space and W a subspace endowed with the same inner product. Then any vector $\boldsymbol{v} \in V$ can be *uniquely* written as $\boldsymbol{v} = \boldsymbol{z} + \boldsymbol{w}$ where $\boldsymbol{w} \in W$ and $\boldsymbol{z}$ is orthogonal to W.

The vector $\boldsymbol{w}$ is called the **orthogonal projection** of $\boldsymbol{v}$ onto the subspace W. Sometimes, it is written as $\text{proj}_W \boldsymbol{v}$. When I say that the vector $\boldsymbol{z}$ is orthogonal to W, I mean that the inner product of $\boldsymbol{z}$ with any vector in W is necessarily 0. The collection of all vectors satisfying this condition is given a special name:

Definition: Suppose V is an inner product space, and W is a subspace of V. The set of all vectors orthogonal to every vector in W is called the **orthogonal complement** of W. The symbol for orthogonal complement is $\perp$; i.e. the orthogonal complement of W is denoted $W^\perp$ and read as "W perp." The "perp" is short for perpendicular.

It is not difficult to show that $W^\perp$ is also a subspace of W; it contains $\mathbf{0}$ and is closed under addition and scalar multiplication by the properties of the inner product. As an example, refer to Fig. 5.5: there, you see that the eigenspaces $\text{Ei}(\lambda = -1)$ and $\text{Ei}(\lambda = 5)$ are orthogonal complements of each other under the usual dot product.

Given an orthogonal basis $\{\boldsymbol{w}_1, \ldots, \boldsymbol{w}_p\}$ for the subspace W, I can write

$$\text{proj}_W \boldsymbol{v} = \sum_{i=1}^{p} \text{proj}_{\boldsymbol{w}_i} \boldsymbol{v}, \tag{6.4.1}$$

since this is necessarily a vector in W by definition of the projection and $\boldsymbol{z} = \boldsymbol{v} - \text{proj}_W \boldsymbol{v}$ is necessarily orthogonal to every vector in $\{\boldsymbol{w}_1, \ldots, \boldsymbol{w}_p\}$ and, hence, in W. Therefore, I have just demonstrated that part of the Orthogonal Decomposition theorem holds: it remains to be seen that this decomposition is *unique.*

To see that $\boldsymbol{z}$ and $\boldsymbol{w} = \text{proj}_W \boldsymbol{v}$ are unique, I must convince you that the only vector common to both $W \neq \{\boldsymbol{0}\}$ and $W^\perp$ is $\boldsymbol{0}$. Suppose $\boldsymbol{u} \in W$ and $\boldsymbol{u} \in W^\perp$. Then, since $\boldsymbol{u}$ is orthogonal to every vector in W and $\boldsymbol{u} \in W$, $\langle \boldsymbol{u}, \boldsymbol{u} \rangle = 0$, and so $\boldsymbol{u} = 0$. It follows that if I can write $\boldsymbol{v} = \boldsymbol{z} + \boldsymbol{w}$ and $\boldsymbol{v} = \boldsymbol{z}' + \boldsymbol{w}'$, where $\boldsymbol{w}, \boldsymbol{w}' \in W$ and $\boldsymbol{z}, \boldsymbol{z}' \in W^\perp$, then

$$\boldsymbol{z} + \boldsymbol{w} = \boldsymbol{z}' + \boldsymbol{w}' \to \boldsymbol{z} - \boldsymbol{z}' = \boldsymbol{w} - \boldsymbol{w}' \in W \cap W^\perp \tag{6.4.2}$$

(the **intersection** of W and $W^\perp$), so $\boldsymbol{w} = \boldsymbol{w}'$ and $\boldsymbol{z} = \boldsymbol{z}'$, proving uniqueness. A corollary of this result is that the particular basis we use to express $\text{proj}_W \boldsymbol{v}$ is inconsequential as long as the basis is orthogonal.

Example: The space $W = \text{Span}\{\cos(t), \sin(t)\}$ is a subspace of $\mathcal{C}([0, \frac{\pi}{2}])$ where I allow for the possibility of *complex*-valued continuous functions. Using the inner product $\langle f, g \rangle = \int_0^{\frac{\pi}{2}} f(t)[g(t)]^* dt$ and the Gram-Schmidt process, it is possible to construct an orthogonal basis for W out of $\{\cos(t), \sin(t)\}$:

$$w_1(t) = \cos(t) \ , \ w_2(t) = \sin(t) - \left(\frac{\langle \sin(t), \cos(t) \rangle}{\langle \cos(t), \cos(t) \rangle} \right) \cos(t) = \sin(t) - \frac{2}{\pi} \cos(t). \tag{6.4.3}$$

The function $1 \in \mathcal{C}([0, \frac{\pi}{2}])$ can be projected onto W:

$$\begin{aligned} \text{proj}_W 1 &= \left(\frac{\langle 1, w_1(t) \rangle}{\langle w_1(t), w_1(t) \rangle} \right) w_1(t) + \left(\frac{\langle 1, w_2(t) \rangle}{\langle w_2(t), w_2(t) \rangle} \right) w_2(t) \\ &= \left(\frac{4}{\pi} \right) \cos(t) + \left(\frac{4}{2 + \pi} \right) \left(\sin(t) - \frac{2}{\pi} \cos(t) \right). \end{aligned} \tag{6.4.4}$$

An alternative basis for W is $\{e^{it}, e^{-it}\}$, which is an immediate consequence of Euler's formula $e^{i\theta} = \cos(\theta) + i\sin(\theta)$. Using the Gram-Schmidt process, I construct from this basis the orthogonal basis

$$u_1(t) = e^{it}\ ,\ u_2(t) = e^{-it} - \left(\frac{\langle e^{-it}, e^{it}\rangle}{\langle e^{it}, e^{it}\rangle}\right) e^{it} = e^{-it} + \frac{2i}{\pi}e^{it}. \tag{6.4.5}$$

Using this basis, I find that

$$\begin{aligned}
\text{proj}_W 1 &= \left(\frac{\langle 1, u_1(t)\rangle}{\langle u_1(t), u_1(t)\rangle}\right) u_1(t) + \left(\frac{\langle 1, u_2(t)\rangle}{\langle u_2(t), u_2(t)\rangle}\right) u_2(t) \\
&= \left(\frac{2(1-i)}{\pi}\right) e^{it} + \left(\frac{2(1+i)}{2+\pi}\right)\left(e^{-it} + \frac{2i}{\pi}e^{it}\right) \\
&= \left(\frac{4}{\pi}\right)\cos(t) + \left(\frac{4}{2+\pi}\right)\left(\sin(t) - \frac{2}{\pi}\cos(t)\right).
\end{aligned} \tag{6.4.6}$$

The third line follows from the second via Euler's formula. So, using two distinct orthogonal bases for W, I computed the projection of 1 and found the same answer independent of the chosen basis. Given that the projection of any vector onto a subspace W must be unique, you should have expected this result from the beginning.

Example: Consider the linearly independent vectors

$$\left\{\vec{v}_1 = \begin{pmatrix} 1 & -1 & 2 \end{pmatrix}^T, \vec{v}_2 = \begin{pmatrix} 1 & 0 & -1 \end{pmatrix}^T\right\}$$

in $\mathbb{R}^3$, and suppose we wish to find the orthogonal decomposition of $\vec{x} = \begin{pmatrix} 1 & 0 & 1 \end{pmatrix}^T$ with respect to the plane W spanned by the two vectors. First, I need to find an orthogonal basis for W using the Gram-Schmidt process:

$$\vec{w}_1 = \begin{pmatrix} 1 & -1 & 2 \end{pmatrix}^T, \vec{w}_2 = \vec{v}_2 - \text{proj}_{\vec{v}_1}\vec{v}_2 = \begin{pmatrix} 7/6 & -1/6 & -4/6 \end{pmatrix}^T. \tag{6.4.7}$$

Since I am free to multiply any basis vector by any nonzero constant that I wish to obtain a new basis, I'll multiply the second vector by 6, so I don't have to deal with pesky fractions;

our new orthogonal basis for W is then

$$\mathcal{B} = \left\{ \vec{u}_1 = \begin{pmatrix} 1 & -1 & 2 \end{pmatrix}^T, \vec{u}_2 = \begin{pmatrix} 7 & -1 & -4 \end{pmatrix}^T \right\}. \tag{6.4.8}$$

The projection of $\vec{x}$ onto W is given by

$$\text{proj}_W \vec{x} = \text{proj}_{\vec{u}_1} \vec{x} + \text{proj}_{\vec{u}_2} \vec{x} = \begin{pmatrix} 9/11 & -6/11 & 9/11 \end{pmatrix}^T. \tag{6.4.9}$$

The results of the Gram-Schmidt process and the projection of $\vec{x}$ can be visualized by examining Fig. 6.3.

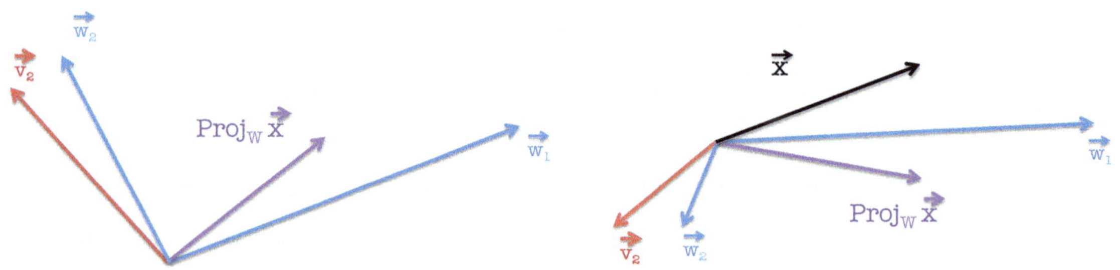

(a) (b)

Figure 6.3: Visualization of the projection of $\vec{x}$ on the subspace $W \subset \mathbb{R}^3$ spanned by $\vec{v}_1 = \vec{w}_1$ (in blue) and $\vec{v}_2$ (in red). Here, $\vec{x}$ is in black and $\text{proj}_W \vec{x}$ is in purple. In figure (a), we are viewing the subspace W (not pictured) from directly below; the vector $\vec{x}$ is not pictured as it is blocked by its projection onto W. Note that the vectors $\vec{w}_1$ and $\vec{w}_2$ are perpendicular as they should be from the Gram-Schmidt construction. In figure (b), we are viewing the subspace W (not pictured) from the side so that $\vec{w}_2$ and $\vec{v}_2$ are coming out of the page, $\vec{w}_1$ points into the page, the projection comes out of the page and points to the right and $\vec{x}$ comes out the page and points to the right, and up.

6.5 The Best Approximation Theorem

I'll examine the last example a little more closely. Any vector in W can be written as a linear combination of $\vec{u}_1$ and $\vec{u}_2$. I don't necessarily need to use the orthogonal basis constructed using Gram-Schmidt for this argument, but generally orthogonal bases are nicer to work with, so I'll use $\{\vec{u}_1, \vec{u}_2\}$ since the components are all integers. If I calculate the square of the distance from $\vec{x}$ to any vector $c_1\vec{u}_1 + c_2\vec{u}_2 \in W$, then I find a function

$$f(c_1, c_2) = \|\vec{x} - (c_1\vec{u}_1 + c_2\vec{u}_2)\|^2, \tag{6.5.1}$$

where the variables are the weights c_1, c_2; as the weights take on all possible values, the linear combination traces out the entire plane W. A direct calculation using the definition of the dot product gives me

$$\begin{aligned} f(c_1, c_2) &= (1 - c_1 - 7c_2)^2 + (c_1 + c_2)^2 + (1 - 2c_1 + 4c_2)^2 \\ &= 2 - 6c_1 + 6c_1^2 - 6c_2 + 66c_2^2. \end{aligned} \tag{6.5.2}$$

It turns out that this function of c_1 and c_2 has a smallest possible value for a unique choice of weights.[4] If I complete the square for both the c_1 and c_2 terms, I find

$$f(c_1, c_2) = 6\left(c_1 - \frac{1}{2}\right)^2 + 66\left(c_2 - \frac{1}{22}\right)^2 + \frac{4}{11}. \tag{6.5.3}$$

This expression has the smallest possible value of $\frac{4}{11}$ precisely when $c_1 = \frac{1}{2}$ and $c_2 = \frac{1}{22}$. These weights correspond to the vector

$$\begin{aligned} c_1\vec{u}_1 + c_2\vec{u}_2 &= \frac{1}{2}\begin{pmatrix} 1 & -1 & 2 \end{pmatrix}^T + \frac{1}{22}\begin{pmatrix} 7 & -1 & -4 \end{pmatrix}^T \\ &= \begin{pmatrix} \frac{9}{11} & -\frac{6}{11} & \frac{9}{11} \end{pmatrix}^T \\ &= \text{proj}_W\vec{x}. \end{aligned} \tag{6.5.4}$$

To summarize, the projection of $\vec{x}$ onto the subspace W is the *unique* vector that achieves the minimum distance to $\vec{x}$ among all of the possible vectors of W! In other words, $\text{proj}_W\vec{x}$ is the best possible representation of $\vec{x} \notin W$ in the subspace W.

Of course, I only demonstrated this for a single example, which does not constitute a proof. But it gets us thinking: is it true in general that the projection of a vector onto a subspace W minimizes the distance using the given inner product? The answer turns out to be yes.

[4]One way to show this is true is by using multivariable calculus ideas. I encourage you to explore this possibility on your own, but I'll take a different approach to keep the calculus minimal.

Let W be a subspace of the inner product space V, and let $\boldsymbol{x} \in V$. From the orthogonal decomposition theorem, we know that there are unique vectors $\boldsymbol{z} \in W^{\perp}$ and $\boldsymbol{w} \in V$ such that $\boldsymbol{x} = \boldsymbol{w} + \boldsymbol{z}$. Let $\boldsymbol{w}'$ be some other vector in W. Then

$$||\boldsymbol{x} - \boldsymbol{w}'||^2 = ||\boldsymbol{z} + (\boldsymbol{w} - \boldsymbol{w}')\,||^2 = ||\boldsymbol{z}||^2 + ||\boldsymbol{w} - \boldsymbol{w}'||^2 \geq ||\boldsymbol{z}||^2 = ||\boldsymbol{x} - \boldsymbol{w}||^2. \tag{6.5.5}$$

This shows that the projection $\boldsymbol{z}$ does indeed make the distance between vectors in W and $\boldsymbol{x}$ as small as possible.

Besides being of geometric interest, this theorem can be used in many approximation problems: that is why mathematicians refer to this fact as the **Best Approximation Theorem.**

Application: Overdetermined Systems

At the very beginning of the book, I mentioned that linear algebra can be applied to CAT scan imagining. In particular, if $\vec{x}$ is a list of the pixel densities obtained from a CAT scan, the image of the interior the person or object can be constructed by solving a linear model of the form $A\vec{x} = \vec{b}$. In this situation, it is typical to have an **overdetermined system**, where the number of equations is larger than the number of variables.

In general, it is not possible to find solutions to $A\vec{x} = \vec{b}$ when the system is overdetermined. However, it is possible to find the best possible *approximate* answer by using the Best Approximation Theorem. Such a solution is called a **least-squares solution**. It is certainly true that if $\vec{b}$ happens to be in Col$A$, then the system necessarily has a solution by definition. However, if $\vec{b}$ is *not* in the column space, then we are free to project $\vec{b}$ onto the subspace Col$A$.

In short, if we can't solve $A\vec{x} = \vec{b}$, we look for solutions to $A\vec{x} = \text{proj}_{\text{ColA}}\vec{b}$ that must exist by definition of the column space. From previous work, we know that a basis for Col$A$ can be obtained by row reducing A and identifying pivot columns; the pivot columns of A form a basis. It is not true in general that this basis is orthogonal, but we can make an orthogonal basis by using the Gram-Schmidt process. So, in principle, we always have an orthogonal basis for Col$A$ at our disposal. Using this orthogonal basis, we can compute the projection of $\vec{b}$ onto the column space.

From the orthogonal decomposition theorem, we can write $\text{proj}_{\text{Col}A}\vec{b}$ as $\vec{b}-\vec{z}$ for some unique $\vec{z} \in \text{Col}A^{\perp}$. Since $\vec{z}$ is perpendicular to every vector in the column space of A, it follows that $A^T\vec{z} = \vec{0}$: multiplying $\vec{z}$ on the left by A^T returns the column vector $\begin{pmatrix} \vec{a}_1 \cdot \vec{z} & \dots & \vec{a}_n \cdot \vec{z} \end{pmatrix}^T$, where $\vec{a}_j$ is the j-th column of A. So, if $\vec{x}$ satisfies $A\vec{x} = \text{proj}_{\text{Col}A}\vec{b} = \vec{b} - \vec{z}$, then it must satisfy the so-called **normal equations:**

$$A^T A\vec{x} = A^T\vec{b}. \tag{6.5.6}$$

Solving the normal equations can be expedited if the matrix A has a **QR**-factorization:

Theorem (QR Factorization): If A is an $m \times n$ matrix with linearly independent columns then A can be factored as $A = QR$ where Q is an $m \times n$ matrix, for which the columns form an orthonormal basis for $\text{Col}(A)$, and R is an $n \times n$ upper triangular invertible matrix with positive entries along the main diagonal.[5]

I would like to illustrate this theorem with an example, since there is much in the way of terminology, and it is easy to get lost in the abstraction. Consider the matrix

$$A = \begin{pmatrix} 1 & 0 & 0 \\ 1 & 2 & 0 \\ 1 & 2 & 3 \\ 1 & 2 & 3 \end{pmatrix} \tag{6.5.7}$$

A straightforward application of the Gauss-Jordan algorithm demonstrates that the columns of this matrix are linearly independent. Therefore, all three columns of A together form a basis for the column space. However, it is easy to see that the vectors are not orthogonal. I can use the Gram-Schmidt process to construct an ortho*normal*[6] basis from the given columns:

$$\left\{ \begin{pmatrix} \frac{1}{2} \\ \frac{1}{2} \\ \frac{1}{2} \\ \frac{1}{2} \end{pmatrix}, \begin{pmatrix} \frac{-3}{\sqrt{12}} \\ \frac{1}{\sqrt{12}} \\ \frac{1}{\sqrt{12}} \\ \frac{1}{\sqrt{12}} \end{pmatrix}, \begin{pmatrix} 0 \\ -\frac{2}{\sqrt{6}} \\ \frac{1}{\sqrt{6}} \\ \frac{1}{\sqrt{6}} \end{pmatrix} \right\} \tag{6.5.8}$$

[5]The statement of the theorem appears here as it does in [5].

[6]Notice that I have taken the liberty of normalizing all the vectors for convenience.

Clearly, if I put these vectors together into a matrix called Q, the columns of Q form an orthonormal basis for the column space of A. Since the columns of Q are orthonormal, a direct calculation shows that $Q^T Q = I_{3\times 3}$: if $A = QR$ as the theorem suggests, then it follows that $Q^T A = \left(Q^T Q\right) R = R$;

$$R = \begin{pmatrix} 2 & 3 & 3 \\ 0 & \sqrt{3} & \sqrt{3} \\ 0 & 0 & \sqrt{6} \end{pmatrix}. \tag{6.5.9}$$

Notice that R is indeed upper triangular and the entries along the main diagonal are positive.

This example should (a) illustrate the general process of finding the QR-factorization of a matrix when the columns of A are linearly independent and (b) the technique for proving the general theorem: given that the columns of A are linearly independent, we can use the Gram-Schmidt process to compute an orthonormal basis for Col$A$. With this basis, we construct Q and define R to be $Q^T A$.

Returning to our original problem, I want to solve the normal equations $A^T A\vec{x} = A^T\vec{b}$. If I assume that A has linearly independent columns, then it is possible to write $A = QR$ as stated in the theorem. From the basic properties of transposes, $A^T = R^T Q^T$, so

$$A^T A\vec{x} = \left(R^T Q^T\right)(QR)\,\vec{x} = R^T R\vec{x} = R^T Q^T \vec{b}. \tag{6.5.10}$$

Since R is upper triangular and the entries along the main diagonal are positive, the determinant of R is given by the product of these diagonal entries, and this product is necessarily positive. Therefore, $\det R \neq 0$, so R^{-1} and hence, $\left(R^T\right)^{-1}$ exists! I can conclude that if $\vec{x}$ solves the normal equations, then $\vec{x} = R^{-1} Q^T \vec{b}$.

Example: It is a common problem to be given a collection of data points and be required to find a best fit model of some desired type. While this is a large and important subject, I only have space to give a simple example illustrating the utility of the normal equations. Suppose we want the best linear model $y = mx + b$ representing the data $(-1, 5), (4, 3), (6, -7), (2, 1), (0, 3)$. As we know, exactly two points are needed to specify the line, specifically the slope and the y-intercept. However, we are given more than two points, and we, therefore, expect the system of equations determining the unknowns m and b to be overdetermined. In fact, if I plug the coordinates of the data points into the linear model, then I find a system of equations $5 = m(-1) + b, 3 = m(4) + b$, etc. that can be written in matrix notation as:

$$\underbrace{\begin{pmatrix} 1 & -1 \\ 1 & 4 \\ 1 & 6 \\ 1 & 2 \\ 1 & 0 \end{pmatrix}}_{A} \underbrace{\begin{pmatrix} b \\ m \end{pmatrix}}_{\vec{x}} = \underbrace{\begin{pmatrix} 5 \\ 3 \\ -7 \\ 1 \\ 3 \end{pmatrix}}_{\vec{b}}. \tag{6.5.11}$$

It is not hard to see that the columns of A are linearly independent: the first column cannot possibly be a multiple of the second. Therefore, it is possible to find a QR-factorization of A. For simplicity, I'll define $\vec{v}_1 = \begin{pmatrix} 1 & 1 & 1 & 1 & 1 \end{pmatrix}^T$ and $\vec{v}_2 = \begin{pmatrix} -1 & 4 & 6 & 2 & 0 \end{pmatrix}^T$. Using the Gram-Schmidt process, an orthogonal basis can be obtained by letting $\vec{w}_1 = \vec{v}_1$ and $\vec{w}_2 = \vec{v}_2 - \text{proj}_{\vec{v}_1}\vec{v}_2$; i.e. $\vec{w}_1 = \begin{pmatrix} 1 & 1 & 1 & 1 & 1 \end{pmatrix}^T$ and $\vec{w}_2 = \begin{pmatrix} -16/5 & 9/5 & 19/5 & -1/5 & -11/5 \end{pmatrix}^T$.

To find Q, it is necessary to have an orthonormal basis for the column space of A. After normalizing $\vec{w}_1$ and $\vec{w}_2$, we have $\vec{u}_1 = \begin{pmatrix} 1/\sqrt{5} & 1/\sqrt{5} & 1/\sqrt{5} & 1/\sqrt{5} & 1/\sqrt{5} \end{pmatrix}^T$ and $\vec{u}_2 = \begin{pmatrix} -16/\left(2\sqrt{205}\right) & 9/\left(2\sqrt{205}\right) & 19/\left(2\sqrt{205}\right) & -1/\left(2\sqrt{205}\right) & -11/\left(2\sqrt{205}\right) \end{pmatrix}^T$. The vectors $\vec{u}_1$ and $\vec{u}_2$ form the columns of Q:

$$Q = \begin{pmatrix} \frac{1}{\sqrt{5}} & \frac{-16}{2\sqrt{205}} \\ \frac{1}{\sqrt{5}} & \frac{9}{2\sqrt{205}} \\ \frac{1}{\sqrt{5}} & \frac{19}{2\sqrt{205}} \\ \frac{1}{\sqrt{5}} & \frac{-1}{2\sqrt{205}} \\ \frac{1}{\sqrt{5}} & \frac{-11}{2\sqrt{205}} \end{pmatrix}, \quad Q^T A = R = \begin{pmatrix} \sqrt{5} & \frac{11}{\sqrt{5}} \\ 0 & 2\sqrt{\frac{41}{5}} \end{pmatrix}. \tag{6.5.12}$$

The inverse of R can easily be calculated: $R^{-1} = \begin{pmatrix} \frac{1}{\sqrt{5}} & -\frac{11}{2\sqrt{205}} \\ 0 & \frac{1}{2}\sqrt{\frac{5}{41}} \end{pmatrix}$. Therefore, upon solving the normal equations, I find the approximate values for m and b:

$$\begin{pmatrix} b \\ m \end{pmatrix} \approx R^{-1}Q^T \begin{pmatrix} 5 \\ 3 \\ -7 \\ 1 \\ 3 \end{pmatrix} = \begin{pmatrix} \frac{162}{41} \\ -\frac{55}{41} \end{pmatrix}. \tag{6.5.13}$$

The best fit line is plotted in Fig. 6.4 along with the given data points. Notice that the best fit line does indeed do a good job of coming as close to each of the data points as possible.

Of course, data won't always be best represented by a linear function. It is possible to use the best approximation theorem to find least-squares fitting using curves besides lines. Though I won't cover this fascinating topic, I encourage the reader to read the references listed at the end of this chapter for more information.

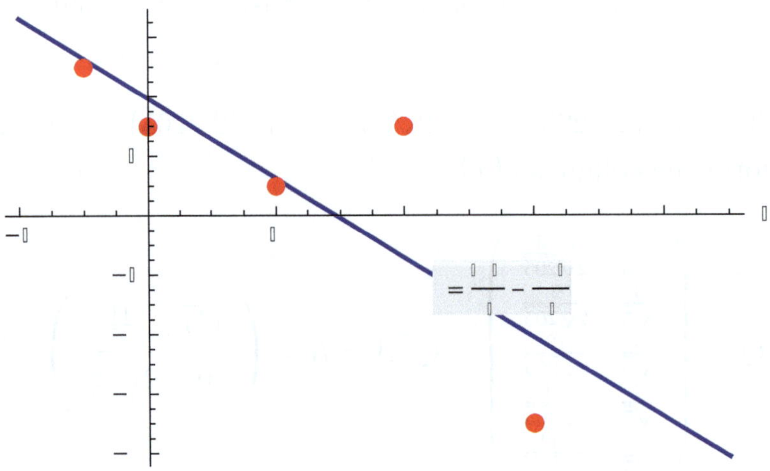

Figure 6.4: The red dots correspond to the data points given in the example. The blue line represents the best fit linear function obtained by solving the normal equations.

Fourier Series

The subject of Fourier series is a cornerstone of applied mathematics. It is applied to the study of certain families of differential equations and provides indispensable tools for signal analysis and physics. Given a function $f(t) \in \mathcal{C}([0,2\pi])$ with the standard inner product, mathematicians define the **order n Fourier approximation of f** to be the projection of $f(t)$ onto the subspace $W = \text{Span}\{1, \cos(t), \cos(2t), \ldots, \cos(nt), \sin(t), \sin(2t), \ldots, \sin(nt)\}$ where n is a positive integer. Fortunately, the functions in the spanning set for W are already orthogonal, so you don't need to apply the Gram-Schmidt process to find an orthogonal basis.

To convince you of this fact, consider the integral

$$\int_0^{2\pi} \sin(mt)\cos(nt)dt, \tag{6.5.14}$$

where m, n are positive integers. If the spanning set for W is indeed orthogonal, then it should be the case that this integral vanishes. One way of computing the integral would be to use a trigonometric identity. However, there is another way that is often more convenient for many mathematical applications. By Euler's formula, we know that $e^{i\theta} = \cos(\theta) + i\sin(\theta)$ holds as an identity. As before, we can use this formula to write $\cos(\theta) = \left(e^{i\theta} + e^{-i\theta}\right)/2,\ \sin(\theta) = \left(e^{i\theta} - e^{-i\theta}\right)/(2i)$. Therefore,

$$\sin(mt)\cos(nt) = \left(\frac{e^{imt} - e^{-imt}}{2i}\right)\left(\frac{e^{int} + e^{-int}}{2}\right) = \frac{e^{i(m+n)t} + e^{i(m-n)t} - e^{i(n-m)t} - e^{-i(m+n)t}}{4i}. \tag{6.5.15}$$

The integrals of the complex exponentials can be done easily using the fundamental theorem of calculus together with $\int e^{ax}dx = \frac{1}{a}e^{ax} + C$ to show that $\int_0^{2\pi} \sin(mt)\cos(nt)dt$ is indeed 0 for any choice of integers m and n. A similar exercise should prove to you that spanning set for W forms an orthogonal basis for W.

Example: I will compute for you the n-th order Fourier approximation for the functions $f(t) = t$ and $f(t) = t\,(2\pi - t)$, both of which belong to $\mathcal{C}([0,2\pi])$. In order to evaluate the relevant integrals on your own, you'll need to use a technique from calculus known as the method of integration by parts.[7] I don't assume you've seen this before, so I'll just record the answers below.

By definition of the n-th order Fourier approximation, I need to compute $\text{proj}_W f(t) = \sum_{j=0}^{n} \text{proj}_{\cos(jt)} f(t) + \sum_{j=1}^{n} \text{proj}_{\sin(jt)} f(t)$, since the best approximation of $f(t)$ in W is the projection of $f(t)$ onto W according to the Best Approximation Theorem. Also by definition,

$$\text{proj}_{\cos(jt)} f(t) = \frac{\langle f(t), \cos(jt) \rangle}{\langle \cos(jt), \cos(jt) \rangle} \cos(jt) = \left(\frac{\int_0^{2\pi} f(t) \cos(jt) dt}{\int_0^{2\pi} \cos^2(jt) dt} \right) \cos(jt) \tag{6.5.16}$$

$$\text{proj}_{\sin(jt)} f(t) = \frac{\langle f(t), \sin(jt) \rangle}{\langle \sin(jt), \sin(jt) \rangle} \sin(jt) = \left(\frac{\int_0^{2\pi} f(t) \sin(jt) dt}{\int_0^{2\pi} \sin^2(jt) dt} \right) \sin(jt). \tag{6.5.17}$$

Computing the relevant integrals using integration by parts, I find

$$f(t) = t \approx \pi + \sum_{j=1}^{n} \left(\frac{-2}{j} \right) \sin(jt). \tag{6.5.18}$$

To visualize the approximation, I choose a handful of n values and plotted the approximation with $f(t) = t$ on the same coordinate plane. Referring to Fig. 6.5, you see that the Fourier approximations tend to improve as n increases. This is generally true under relatively mild assumptions about $f(t)$. However, notice that the approximations fail to do a good job in approximating $f(t) = t$ near the endpoints 0 and 2π. This is known as the **Gibbs Phenomenon** and has to do with the fact that the functions we are using to approximate $f(t)$ are continuous and periodic in t whereas $f(t)$ is not; in fact, if I extend $f(t) = t$ to be 2π- periodic for all t, then we have a discontinuous function for which the graph resembles the teeth on a saw. (See Fig. 6.6.) In contrast to this example, the n-th order Fourier approximation of $f(t) = t\,(2\pi - t)$ is given by

$$f(t) = t\,(2\pi - t) \approx \frac{2}{3}\pi^2 - \sum_{j=1}^{n} \frac{4}{j^2} \cos(jt). \tag{6.5.19}$$

[7]In short, integration by parts relies on the integral formula $\int_a^b f(x)g'(x)dx = f(b)g(b) - f(a)g(a) - \int_a^b f'(x)g(x)dx$.

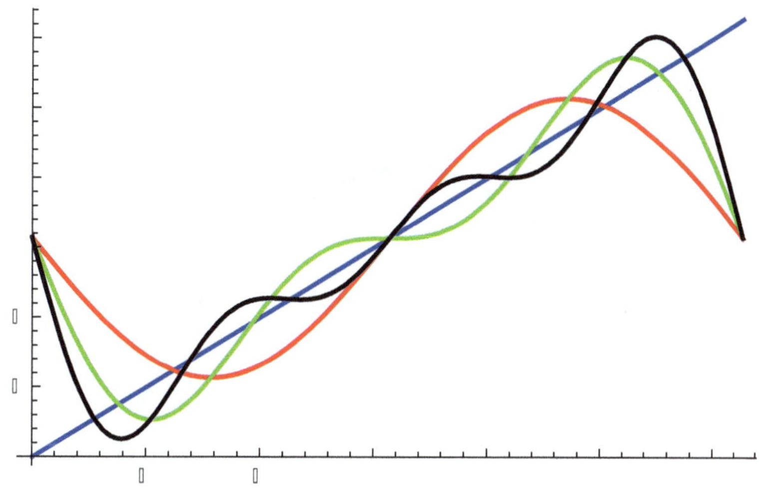

Figure 6.5: Plot of $f(t) = t$ on $[0, 2\pi]$ (blue) together with the $n = 1$ (red), $n = 2$ (green), and $n = 3$ (black) Fourier approximations of $f(t)$.

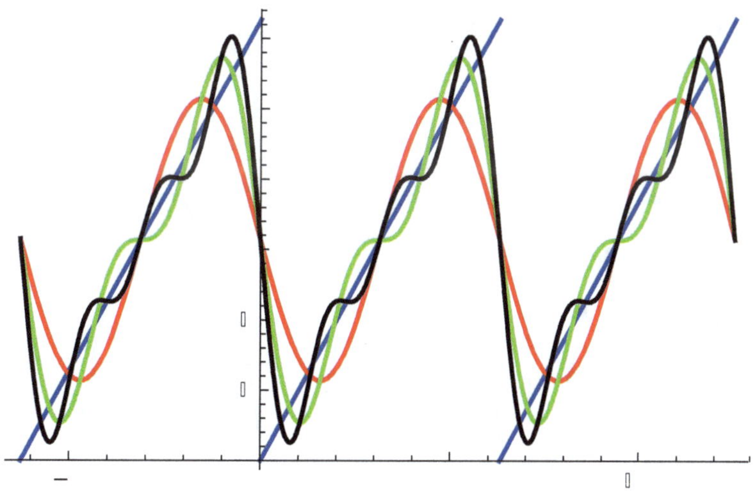

Figure 6.6: Plot of $f(t) = t \bmod 2\pi$ on $[-2\pi, 4\pi]$ (blue) together with the $n = 1$ (red), $n = 2$ (green), and $n = 3$ (black) Fourier approximations of $f(t)$. The Fourier approximations are always continuous at the discontinuities of $f(t) = t \bmod 2\pi$.

I plot the first few Fourier approximations of $f(t)$ together with the function itself in Fig. 6.7; in Fig. 6.8, I plot the approximations on a larger interval together with the 2π-periodic extension of $f(t)$. Notice that this extension is continuous everywhere, and the approximations behave better near corner points of the extension.

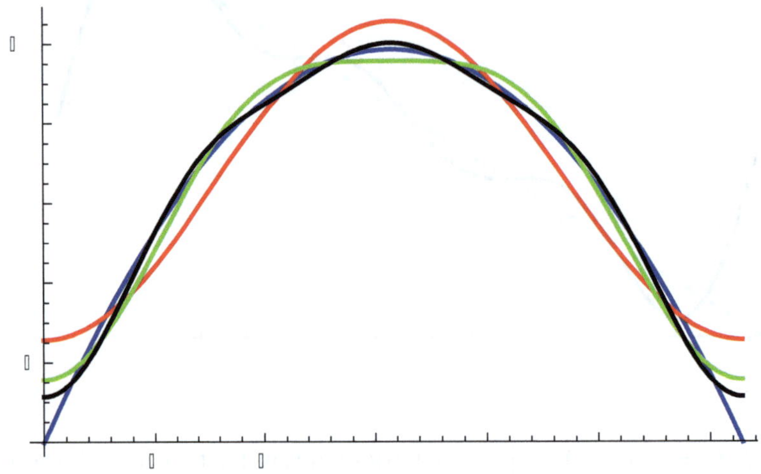

Figure 6.7: Plot of $f(t) = t(2\pi - t)$ on $[0, 2\pi]$ (blue) together with the $n = 1$ (red), $n = 2$ (green) and $n = 3$ (black) Fourier approximations of $f(t)$.

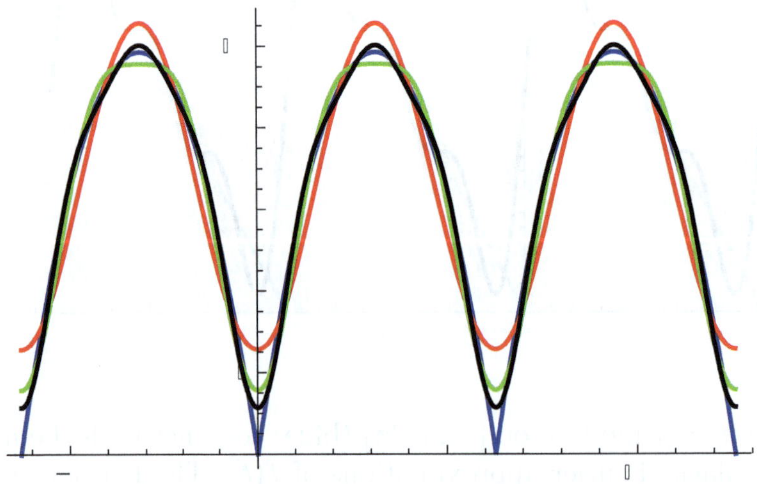

Figure 6.8: Plot of $f(t) = (t \bmod 2\pi)\,(2\pi - t \bmod 2\pi)$ on $[-2\pi, 4\pi]$ (blue) together with the $n = 1$ (red) , $n = 2$ (green), and $n = 3$ (black) Fourier approximations of $f(t)$.

To finish this section, I would like to give you a glimpse as to how Fourier series can be used to solve partial differential equations in physics. Earlier in the book, we looked at a basic example of such an equation, namely the heat equation. There, we were interested in finding the steady-state temperature distribution on a rectangular plate by numerically solving the heat equation. I'll discuss now another partial differential equation that describes how waves move. Suppose we have a flexible elastic string attached to $x = 0$ and $x = 2\pi$ along the x-axis. If we define $u(x,t)$ to be the height of the string above the x-axis (or below if u is negative) at point x at time t, then $u(x,t)$ is governed by the equation

$$\frac{1}{c^2}\frac{\partial^2 u}{\partial t^2} = \frac{\partial^2 u}{\partial x^2}, \tag{6.5.20}$$

called the **one-dimensional wave equation**. Here, c is just a constant with units of speed. sine and cosine functions are the perfect models for wave phenomena (think of their graphs!), so we expect that Fourier approximations will be useful here. For sake of argument, let's say that I hold the string so that the shape of the string at $t = 0$ is given by $u(x,0) = \sin(x)$. Don't worry about the units of t and x; I want to keep the math as simple as possible. My mission, then, is to solve the wave equation with this **initial data**. Once we let go of the string, it will vibrate according to the wave equation.

This is a challenging problem, but I know that I can approximate $u(x,t)$ at any fixed t very well by a Fourier approximation: $u(x,t) \approx \sum_{j=1}^{n} A_j(t)\sin(jt)$. The larger the value of n, the better the approximation. Here, I'm assuming that the coefficients are functions of time to be determined. Notice that there are no cosine terms, because I insist that the string remain fixed in place at the endpoints and only the sine functions vanish at $x = 0$ and $x = 2\pi$. If I plug this approximation into the wave equation, I find (setting $c = 1$):

$$\sum_{j=1}^{n} A_j''(t)\sin(jx) = \sum_{j=1}^{n} A_j(t)\left(-j^2\right)\sin(jx). \tag{6.5.21}$$

Since the sine functions are orthogonal and hence linearly independent, it must be true that $A_j''(t) = -j^2 A_j(t)$. Also, I must require that $A_1(0) = 1$ and $A_j(0) = 0$ $(j \neq 1)$, since the initial shape of the string is given by $u(x,0) = \sin(x)$. The unique solution to this set of differential equations is $\{A_1(t) = \cos(t), A_{j\neq 1}(t) = 0\}$, so $u(x,t) \approx \cos(t)\sin(x)$. In fact, this solution is exact as you can verify directly. The plot of $u(x,t)$ is given in Fig. 6.9. Since it is a function of two variables, its graph is a surface in $\mathbb{R}^3$. The cross-section of the surface for any fixed t tells you the shape of the string as a function of x. As t increases from 0, the string vibrates in a way so that the maximum and minimum switch places periodically.

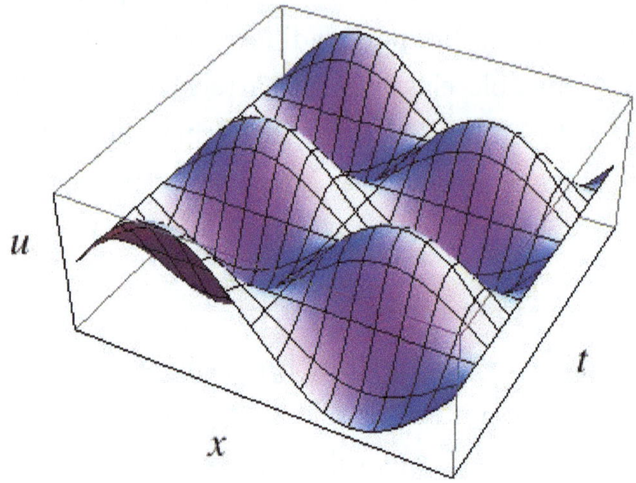

Figure 6.9: Plot of $u(x,t) = \cos(t)\sin(x)$, a solution to the one-dimensional wave equation. For any fixed t, the corresponding cross-section of the surface shows the shape of the string $u(x,t)$ as a function of x. As you move your attention along the t direction, the maximum and minimum of the string switch positions periodically.

This is just one example of the usefulness of Fourier series in physics and partial differential equations. For more information, the reader is encouraged to refer to the books listed at the end of this chapter.

6.6 Chapter 6 Exercises

1. Compute the distance between $\vec{u} = (1, -2, 0, 3)^T$ and $\vec{v} = (0, -2, 5, 1)^T$, and normalize the vectors.

2. Compute the distance between $\vec{u} = (1, -2i, 0, 3+i)^T$ and $\vec{v} = (i-1, -2, 5, i+2)^T$ and normalize the vectors. Use the inner product $\langle \vec{u}, \vec{v} \rangle = \vec{v}^\dagger \vec{u}$ for general complex column vectors $\vec{u}, \vec{v} \in \mathbb{C}^n$.

3. Compute the distance between $\boldsymbol{u} = t^2+2t+1$ and $\boldsymbol{v} = t-1$, and normalize the vectors. Use the inner product $\langle \boldsymbol{u}, \boldsymbol{v} \rangle = \int_{-1}^{1} \boldsymbol{u}(t)\boldsymbol{v}(t)dt$ for vectors in P_2.

4. Compute the distance between $\boldsymbol{u} = t^2+2t+1$ and $\boldsymbol{v} = t-1$, and normalize the vectors. Use the inner product $(\boldsymbol{u}, \boldsymbol{v}) = \boldsymbol{u}(-1)\boldsymbol{v}(-1) + \boldsymbol{u}(0)\boldsymbol{v}(0) + \boldsymbol{u}(1)\boldsymbol{v}(1)$ for vectors in P_2. Compare your results to the previous problem.

5. Find all vectors in $\mathbb{R}^4$ orthogonal to both $\vec{u} = (1, -2, 0, 3)^T$ and $\vec{v} = (0, -2, 5, 1)^T$. If W is the span of $\vec{u} = (1, -2, 0, 3)^T$ and $\vec{v} = (0, -2, 5, 1)^T$, then the set you compute is called the orthogonal complement to W, $W^\perp$, which is a subspace of $\mathbb{R}^4$. What is the dimension of this space?

6. Prove that for any nontrivial subspace W of $\mathbb{R}^n$, $\dim W + \dim W^\perp = n$.

7. Find all vectors in $\mathbb{C}^4$ orthogonal to both $\vec{u} = (1, -2i, 0, 3+i)^T$ and $\vec{v} = (i-1, -2, 5, i+2)^T$. Use the inner product $\langle \vec{u}, \vec{v} \rangle = \vec{v}^\dagger \vec{u}$ for general complex column vectors $\vec{u}, \vec{v} \in \mathbb{C}^n$.

8. Find the orthogonal projection of $\vec{u} = (1, -2, 0, 3)^T$ onto $\vec{v} = (0, -2, 5, 1)^T$.

9. Find the orthogonal projection of $\vec{u} = (1, -2i, 0, 3+i)^T$ onto $\vec{v} = (i-1, -2, 5, i+2)^T$ using the inner product $\langle \vec{u}, \vec{v} \rangle = \vec{v}^\dagger \vec{u}$.

10. Find the orthogonal projection of $\boldsymbol{u} = t^2 + 2t + 1$ and $\boldsymbol{v} = t - 1$ using first the inner product $(\boldsymbol{u}, \boldsymbol{v}) = \boldsymbol{u}(-1)\boldsymbol{v}(-1) + \boldsymbol{u}(0)\boldsymbol{v}(0) + \boldsymbol{u}(1)\boldsymbol{v}(1)$ for vectors in P_2 and then the inner product $\langle \boldsymbol{u}, \boldsymbol{v} \rangle = \int_{-1}^{1} \boldsymbol{u}(t)\boldsymbol{v}(t)dt$. Compare your results.

11. Demonstrate that the Cauchy-Schwarz inequality holds for the vectors

$$(-1, 2+3i, 4)^T, (-2-i, 0, 3+2i)^T$$

using the inner product $\langle \vec{u}, \vec{v} \rangle = \vec{v}^\dagger \vec{u}$.

12. Use the Gram-Schmidt process to find an orthonormal basis for Col$A$ and Nul$A$ where $A = \begin{pmatrix} -5 & 1 & -10 & -15 & -1 \\ 1 & 0 & 2 & 3 & 0 \end{pmatrix}$.

13. Use the Gram-Schmidt process to compute an orthonormal basis for $P_2 = \text{Span}\{1, t, t^2\}$. First, use the inner product $(\boldsymbol{u}, \boldsymbol{v}) = \boldsymbol{u}(-1)\boldsymbol{v}(-1) + \boldsymbol{u}(0)\boldsymbol{v}(0) + \boldsymbol{u}(1)\boldsymbol{v}(1)$ for vectors in P_2 and then the inner product $\langle \boldsymbol{u}, \boldsymbol{v} \rangle = \int_{-1}^{1} \boldsymbol{u}(t)\boldsymbol{v}(t)dt$. Compare your results.

14. Use the Gram-Schmidt process to find an orthonormal basis for $\mathbb{R}^3$ consisting of eigenvectors of the symmetric matrix $A = \begin{pmatrix} 0 & 1 & 1 \\ 1 & 0 & 1 \\ 1 & 1 & 0 \end{pmatrix}$. In general, finding an orthogonal basis of eigenvectors for a given symmetric matrix A is referred to as **orthogonal diagonalization**.

15. Demonstrate that the columns of the matrix $A = \begin{pmatrix} 1 & 2 \\ 3 & 4 \\ 0 & -1 \\ -2 & 1 \end{pmatrix}$ are linearly independent. Use this fact and the QR-factorization theorem to determine the best possible solution to $A\vec{x} = \vec{b}$ where $\vec{b} = (3, 7, -1, 3)^T$. Assume the inner product used in this problem is the standard one on $\mathbb{R}^n$.

16. Find the best fit line containing the points $(-1, 4), (3, 5), (7, 1), (-1, 0)$, and $(0, 0)$. Assume the inner product used in this problem is the standard one on $\mathbb{R}^n$.

17. Compute the $n = 4$ Fourier approximation to $f(t) = |t - 2\pi|$.

18. Consider the inner product space $\mathcal{C}([-1, 1])$ consisting of all continuous functions on the interval $[-1, 1]$ with inner product $\langle \boldsymbol{u}, \boldsymbol{v} \rangle = \int_{-1}^{1} \boldsymbol{u}(t)\boldsymbol{v}(t)dt$. Define W to be the subset of all even functions, where a function is even if it satisfies $f(t) = f(-t)$. Show that W is a vector subspace, and then, compute the orthogonal complement $W^\perp$. Show that it is always possible to write a function in $\mathcal{C}([-1, 1])$ as the sum of an element from W and an element from $W^\perp$.

6.7 Chapter 6 References

If the reader wishes to explore the theory behind the theorems presented in this chapter or wishes to see more examples, I suggest consulting [5, 6]. Readers interested in learning more about Fourier series should consult [11]. The book [5] has a more extensive discussion of least squares regression in the context of linear algebra.

Chapter 7

Complex Numbers

You have probably heard it said that it is not possible to take the square root of a negative number. What is meant by that statement is that there is no real number such that $y = \sqrt{x}$ where $x < 0$ since then $y^2 = x < 0$; if y were real, then $y^2 \geq 0$, necessarily. However, there is a multitude of situations where we must make sense of square roots of negative real numbers. Mathematics is quite incomplete without so-called **complex numbers**:

Definition: A **complex number** is a number of the form $z = a + bi$, where $a, b \in \mathbb{R}$. Mathematicians define the **imaginary** number $i = \sqrt{-1}$ such that $i^2 + 1 = 0$.

Just like real numbers, we can define addition and multiplication of complex numbers: suppose $z = a + bi$ and $w = c + di$. Then

$$z + w = (a + c) + (b + d)i \ , \ \ zw = (ac - bd) + (ad + bc)i. \tag{7.0.1}$$

Definition: If $z = a + bi$, then a is called the **real part** of z, and b is called the **imaginary part**. Often times, the symbol Re z is used for the real part, and Im z is used for the imaginary part.

Complex numbers may be represented in the **complex plane**, which is just a copy of $\mathbb{R}^2$. The x-axis is called the real axis, and the y-axis is called the imaginary axis. The relationship between a complex number and a point in the plane is given by $a + bi \leftrightarrow (a, b)$. For every nonzero complex number, there is a mirror image of this number across the real axis called the **conjugate**:

Definition: The **conjugate** of z is defined by $z^* = a - bi$.

Notice that $zz^* = a^2 + b^2$, which is the square of the distance from the complex number to the origin of the complex plane. One of the most useful formulas involving complex numbers is **Euler's formula**, which I will state without proof:

$$e^{i\theta} = \cos\theta + i\sin\theta. \tag{7.0.2}$$

This amazing formula comes up in a wide variety of physics and engineering applications and relates two elementary (but seemingly very different) types of functions, namely the exponential functions and the trigonometric functions. By taking the complex conjugate of Euler's formula, we are able to write sine and cosine in terms of complex exponentials:

$$\cos(\theta) = \frac{e^{i\theta} + e^{-i\theta}}{2}, \quad \sin(\theta) = \frac{e^{i\theta} - e^{-i\theta}}{2i}. \tag{7.0.3}$$

These expressions allow us to compute integrals involving cosine and sine using tricks specific to exponential functions. Moreover, these expressions may be used to prove a wealth of trigonometric identities simply based on the laws of exponents.

Bibliography

[1] Koblitz, N., *A Course in Number Theory and Cryptography*, 2nd edition, Springer-Verlag Graduate Texts in Mathematics, 1994.

[2] Chartrand, G., *Introductory Graph Theory*, Dover Publications 1977.

[3] Burden, R.L. and Faires, J.D., *Numerical Analysis*, 10th edition, Brooks/Cole, 2015.

[4] Hamsapriye, *Role of Linear Algebra and its Applications in Health Care Monitoring*, IJITR, Volume 2, Pages 832-836, 2014.

[5] Lay D.C., Lay S.R. and McDonald J.J., *Linear Algebra and Its Applications*, 5th edition, Pearson, 2015.

[6] Hoffman K. and Kunze R., *Linear Algebra*, 2nd edition, Prentice Hall, 1971.

[7] Yuster, T., *The Reduced Row Echelon Form of a Matrix is Unique: A Simple Proof*, Mathematics Magazine, Volume 57, Number 2, Pages 93-94, March, 1984.

[8] D'Angelo J.P. and West D.B., *Mathematical Thinking: Problem-Solving and Proofs*, 2nd edition, Pearson, 1999.

[9] Rosen, K. *Discrete Mathematics and Its Applications*, 6th edition, McGraw-Hill Education, 2006.

[10] Boyce W.E. and DiPrima, R.C., *Elementary Differential Equations and Boundary Value Problems*, 10th edition, Wiley, 2012.

[11] Wilcox, H.J. and Myers, D.L., *An Introduction to Lebesgue Integration and Fourier Series*, Dover Publications, 1994.

Printed by Libri Plureos GmbH in Hamburg,
Germany

9 781516 502554